H.S. AKRONGOLD

GRAND UNIFICATION
OF A
2 PHOTON UNIVERSE

QUANTUM SCIENCE BOOKS
NEW YORK

GRAND UNIFICATION
OF A
2 PHOTON UNIVERSE

A Grand Unified Theory of Everything!

With elegant simplicity, the author has resolved the wave/particle duality paradox by combining single **wave-propertied photons** and single **particle-propertied photons** to construct diagrams of all standard model-entities identified in the framework of modern physics (proton, neutron, antiproton, quark, antiquark, gluon, antigluon, neutrino, electron, W+, W- and Z^0 bosons etc.)

Thereafter, reaction diagrams illustrate the **step-by-step synthesis** of each of these entities, as well as their **photon content** and **spatial arrangement**. Also, for the first time, the unique photon units which generate each of the necessary **forces** (strong, weak, electromagnetic) to produce these steps are identified and diagramed as to their photon composition and spatial structure. For example, the unit generating **the strong force, the gluon/antigluon meson**, is diagramed; as are detailed sequences showing how the meson organizes respective quarks and antiquarks into protons, neutrons and their antiparticles; as well as diagrams demonstrating how and what each antiparticle was thereafter converted into.

From the **Singularity** through a step-by-step synthesis and grand unification of all the model-entities in the framework of standard physics; as well as the forces, fields and bonds required to create them; using a **55 dimensional system** as the template and 2 distinctly different types of photons, a particle type and a wave type, to effectuate the creation of the existent universe. Unique diagrams are presented, using the two photon types, to illustrate the composition and structure of : (+) and (-) monopoles; W+, W-, and Z^0 bosons; electron; positron; neutrino and antineutrino; gluon and antigluon; up and down quarks, and their antiparticles; the proton and neutron and their antiparticles; all of the isotopes of Hydrogen, Helium and Lithium; as well as the composition and structure of the aforementioned forces. Also, how each of **these forces** is actually a part of the **gravity force**, and thus **unified** with it.

Additionally, a unique **double-quark** unit and structure is diagramed, which illustrates the mechanism of how quarks on protons and neutrons of elements are **spliced together** to effectuate the fusion of smaller and lighter elements into larger and heavier elements of the periodic table.

Another presentation illustrates how **two types of gravitons** are the actual precursors of the two types of photons; as well as how they direct the numerous reactions producing each of the particle-entities found in modern physics.

GRAND UNIFICATION OF A 2 PHOTON UNIVERSE

For further information, contact:

Quantum Science Books LLC.
99 Spring Street,
3rd FL,
New York, NY 10012
www.grandunificationscience.com

ISBN: 151712043
LCCN: TXU1-247-135

Dedicated to:
W.D.KENTON

With unending gratitude to:

ROCHELLE STARK
FRANK ARACA
JAMES WALLS WHITE
BARRY AND BRUCE
ISAAC BENJAMIN

Contents

Contents (continued)

Contents (continued)

Part III

Contents (continued)

PROLOGUE

The jet hovered at 36,000 feet. Below, a sea of white pillow-puffs almost hid the black picket fence the Rocky Mountains had erected between the Pacific Ocean and the Great Plains.

I had just finished reading the final sentences of a book by W. D. Kenton. In it, he outlined a fifty-five dimensional structure, in a special technical language, whose associated subject matter was seemingly unrelated to the sciences. Yet, the structure appeared to hold out the possibility of unifying all of the standard model-entities, forces, fields and bonds found in the framework of modern physics. However, after an hour of frustration at not being able to devise a coherent theory as to how to do this, I began to fumble through some other publications I had brought along.

One that caught my eye had an article on photon experiments. After reading it, I was quite exasperated by its conclusion that a single photon had both the properties of a particle and a wave. This, in my mind, was "scientific magic". It simply could not be. I almost consigned the periodical to a nearby trash heap when an insight flickered, and I re-read the article.

This time I concluded that the results of the carefully crafted and painstakingly executed experiments could just as easily be read as there being two distinctly different photons present and detected, one a particle-propertied photon and the other a wave-propertied photon. Much later, I discovered that a perpetual size-ratio difference between the minute wave photon, $(1.6 \times 10^{-33}\text{cm})^3$, and the relatively immense particle photon, $(1.6 \times 10^{-33}\text{cm})^3 \times (9 \times 10^{20}\text{cm.})$, was the cause of the inability to detect and differentiate between the two types of photons.

At that moment, however, another insight began bubbling through. These two types of photons and the fifty-five dimensional structure could be combined to produce the actual structure of the standard model-entities in physics.

Magnetic monopoles, W+, W- and Z^0 bosons, electrons, positrons, quarks and antiquarks, gluons and antigluons, neutrinos and antineutrinos, protons, neutrons, antiprotons and antineutrons eventually would evolve from combining these two different photon types. This will be discussed herein.

In addition, a new understanding developed as to how gravity and the standard model-entities, forces, fields and bonds are unified, as well as an expanded conception of energy and mass-matter in the $E=MC^2$ relationship.

In order to present all of the above in a coherent manner, the presentation will involve an exposition of five premises. The First two will be the subject matter of Part I., which outlines in precise detail all of the standard entities in physics, including diagrams of their structure; as well as the reaction diagrams showing how they were produced.

Part III. deals with the final three premises, including an exposition of the 55 dimensional structure of the existent Universe and the seven epochal stages of the synthesis of all the elements, as well as the structuring of our solar system.

The linchpin will be Part II, dealing with the pre-existent universe and the "Singularity" as the pre-requisite for the step-by-step evolvement of a fifty-five dimensional universe and all of the standard model-entities outlined in the present framework of physics; as well as the forces, fields and bonds that are brought into play in order to forge them into existence, thereby achieving grand unification.

Finally, the epilogue, which outlines the evolution of the sentient Universe and it's ultimate entity, the homo-sapiens; including an understanding of why they are the only entities capable of being conscious of being conscious.

Precis : For pages 11 to 41

A single photon universe, populated by only one type of photon with two properties, particle and wave, has lead to a dead-end for nearly a hundred years in identifying the composition and structure of all entities found therein by physicists.

In the following thirty pages, however, using two separate types of photons: 1.) a particle-type photon, P; and 2.) a wave type photon, V; all of the known-entities: monopoles, bosons, electron, quarks, etc., will be identified as to composition and structure, as well as other properties such as charge, electromagnetic field and force.

A particle-propertied photon (P) and a wave-propertied photon (V) are the only types of photons in the existent Universe; composing all of the entities therein, except for two types of gravitons: the P_E graviton and the V_E graviton; which create each type of photon from their own respective, unique substance.

The P photon is relatively immense in spatial volume: $[(1.6 \times 10^{-33}\text{CM})]^3 \times [(9 \times 10^{20}\text{CM})]$, which is the Planck length cubed, multiplied by the number value of the C^2 constant; this latter number value also being the expansion factor of the Universe.

The V photon, however, has a minute spatial volume of $[(1.6 \times 10^{-33}\text{CM})]^3$, the Planck length cubed; a minute value, just above what Quantum Mechanic's definition of nothingness is, which seems to explain why it has escaped detection as a second type of photon. It is either hidden by P,or undetectable due to its size.

The (P) photon, exhibits a charge of $^1/_{12}$ of that of the electron, (-1), or (-) $^1/_{12}$; while the (V) photon's (+) $^1/_{12}$ charge is $^1/_{12}$ of the charge of a positron or a proton, (+) 1.

With these two types of photons, all of the model-entities known to physics (protons, neutrons, electrons, quarks, gluons, etc.) can and will be constructed and diagrammed herein; as well as then being able to be reacted; resulting in their **reaction products** always having the exact same number of P and V photons as their **reactants** had. No photon is lost in any of these reactions.

Two (P) photons can merge together, producing the (-) monopole, $(PP)^{(-)1/6}$; while two V photons can merge to produce the (+) monopole, $(VV)^{(+)1/6}$; due to both being in a magnetic dimension, where the rule is like attracts and bonds, while opposites repel.

Each type of monopole, still in a magnetic dimension, will then merge with a like monopole, respectively creating a W-boson, (PP/PP)$^{(-)1/3}$, which is a unit of electromagnetic field; as well as a W+ boson, (VV/VV)$^{(+)1/3}$, the unit of electromagnetic force.

Merging three of each of these two respective types of bosons, produces the electron, 3[(W-)]$^{(-)1}$, [(PP/PP)(PP/PP)(PP/PP)]$^{(-1)}$; and the positron [3(W+)]$^{(+)1}$, [(VV/VV)(VV/VV)(VV/VV)]$^{(+1)}$. A reaction then ensues between the electron and the positron, which results in eight different reaction products; as well as a third type of boson, the Z, (PP/VV)^{0}. The eight reaction products are:

1. Neutrino, [3(Z)]^{0}; [(VV/PP)(VV/PP)(VV/PP)]^{0}.
2. Antineutrino, [(W-)(Z)(W+)]^{0}; [(PP/PP)(VV/PP)(VV/VV)]^{0}.
3. Down quark, [2(Z)(W-)]$^{(-)1/3}$; [(VV/PP)(VV/PP)(PP/PP)] $^{(-)1/3}$
4. Antidown quark, [2(Z)(W+)]$^{(+)1/3}$; [(VV/PP)(VV/PP)(VV/VV)] $^{(+)1/3}$
5. Upquark, [(Z)2(W+)]$^{(+)2/3}$; [(VV/PP)(VV/VV)(VV/VV)] $^{(+)2/3}$
6. Antiup quark, [(Z)2(W-)]$^{(-)2/3}$; [(VV/PP)(PP/PP)(PP/PP)] $^{(-)2/3}$
7. Gluon, [2(W-)(W+)]$^{(-)1/3}$; [(PP/PP)(PP/PP)(VV/VV)] $^{(-)1/3}$
8. Antigluon, [2(W+)(W-)]$^{+)1/3}$; [(VV/VV)(VV/VV)(PP/PP)] $^{(+)1/3}$

The latter two entities (7&8), however, are only stable in a merged form as a gluon/antigluon meson^{0} (g/ā^{0}), which exists in nine different forms, four of which are stable and five are unstable.

The gluon and the antigluon, as well as the four model-entities described above, 3 through 6, due to the multiple positions each type of boson can assume in each of them, can therefore exist in three standard configurations, each of which is arbitrarily assigned a color: red, yellow or blue. These six are the only color-entites.

The positron and the electron, however, can not exhibit any of these three color configurations, due to each having a homogeneous boson content; while the neutrino and antineutrino having a zero (0) charge are also categorized as being colorless.

The above described ten electric phase model-entities: five particle-entities and their five antiparticle-entities, are the minute building-blocks that interact within a unique system outlined herein; which eventually evolve into the five bedrock units which will constitute all of the elements of our universe: two types of protons, two types of neutrons and a single electron.

PART I

First and Second Premises

First Premise

In the existent universe, there are two different types of photons present: (1) a particle-propertied photon, herein called a P photon (P), which exhibits the properties and functions of a particle; and (2) a wave-propertied photon, hereinafter referred to as a V photon (V), which displays the properties and functions associated with that of a wave.

Each P photon, or V photon, is a distinct essence. Both have the same mass; but the former is initially contained in a relatively huge volume of space: $[(1.6 \times 10^{-33}\text{cm.})^3 \times (9 \times 10^{20}\text{cm.})]$; while the latter space is initially quite minute: $[(1.6 \times 10^{-33}\text{cm.})^3$.

Second Premise

The P and/or V photons constitute every entity in the existent universe, except for the P_E and V_E gravitons.

Discussion

The P and V photons are generated in and from the substance of their respective parent gravitational entities, the P_E and V_E gravitons. This will be discussed in much greater detail in Part III.

The newly formed V and P photons are each then bonded together with a like V or a like P photon, respectively, to form two types of entities: the plus (+) monopole (VV) and the minus (-) monopole (PP), respectively.

Plus (+) Monopole (VV)$^{(+)1/6}$

The (+) monopole is a unit formed by two V photons fusing together (VV), forming a (+) photon to (+) photon magnetic bond between them: $(V_+{}^+V\ )$. It displays a charge of (+) 1/6 of that exhibited by a proton, or positron, which is (+) 1.

We identify charges of (+) or (-) 1/6, or less, as magnetic-type charges. All particles with charges above this threshold are considered electric: (+ or -) 1/3, (+ or -) 2/3, (+ or -) 1 and 0. The reason the (VV) unit is referred to as a (+) magnetic monopole is that it behaves as if it were a single isolated (+) pole magnet, or North Pole.

Additionally, the reader will see in the "Positron" section a discussion which illustrates that the positron is composed of twelve V photons; thereby indicating that each V photon must carry a charge of (+) 1/12 in order for the positron to have a (+) 1 charge.

Therefore, it follows that each (+) monopole, (VV), must have a net charge of 2 x (+)1/12, or (+) 1/6.

The (+) 1 charge manifested by the proton, however, is the result of a much more complex structure. Here, twenty-four P photons, each having a (-) 1/12 charge, are combined with thirty-six V photons, each presenting a (+) 1/12 charge, a net excess of 12 V photons, which adds up to a total (+) 1 charge displayed by the proton (12 x (+) 1/12).

Minus (-) Monopole (PP) $^{(-)1/6}$

The minus (-) monopole is created by the bonding together of two P photons to form the (PP) unit. A (-) photon to (-) photon magnetic bond then exists between these two photons: (P=P). The electron, paralleling its opposite entity the positron (also called the anti-electron), will be shown to be composed of twelve P photons, each P thus contributing a (-) 1/12 charge to produce the electron's (-) 1 charge.

The (PP) monopole, having two P photons, thus carries a charge of 2 x (-) 1/12, or (-)1/6. It is referred to as a (-) magnetic monopole due to the fact that it can be interpreted as a single isolated (-) pole magnet, or South Pole.

Bosons: W+, W- and Z^0

If a (VV) and a (PP) are respective (+) and (-) magnetic poles, then combining them together into a unit designated as (VV/PP) or (PP/VV) (the / is the weld or magnetic bond between (VV) and

(PP)) creates a complete magnet with both a (+) and (-) side; or a North and South Pole, respectively. It is called a Z^0 boson. The 0 in the Z^0 indicates a net electric charge of zero; as the (-) 1/6 (PP) and the (+) 1/6 (VV) charges each of the respective sides carries neutralize each other. Hereinafter, we will drop the 0 and just use Z to reference Z^0.

If, on the other hand, we bond together a (PP) to another (PP) we would get a W (-) boson, (PP/PP), which exhibits a net overall electrical charge of (-)1/3 [(-)1/6 + (-)1/6)].

If we bond a (VV) to another (VV), we produce a W+ boson, (VV/VV), manifesting a net overall electrical charge of (+)1/3 [(+)1/6 + (+)1/6)].

<u>W (-) Boson</u>

The W (-) boson is shown below in A.

A. $[^{(-)1/6}\,(PP/PP)^{(-)1/6}]^{(-)1/3}$

It exhibits two magnetic [(-)1/6)] charges, one on each end, as well as a single electric [(-)1/3] charge, simultaneously.

Each end of the W- boson's (PP/PP) inner core presents a (-)1/6 magnetic-type charge, as shown below in B.

B. $^{1/6(-)}(PP/PP)^{(-)1/6}$

Simultaneously, the W- boson also displays a net (-) 1/3 electrical charge, which is positioned outside the inner magnetic core, appearing and functioning like a halo, as seen in A above.

This unique type of charge and photonic structutre, a(-)1/3 electric charge surrounding an inner core of twin magnetic charges of (-) 1/6 at each of the ends, is herein defined as <u>a unit of electromagnetic field</u>; which in excess multiples produces (-) entities.

<u>W (+) Boson</u>

The W (+) boson parallels the W (-) boson. It is depicted below in C.

C. $[^{(+)1/6}(VV/VV)(+)^{1/6}]^{(+)1/3}$

It also has a magnetic charged inner core, as presented in D below.

D. $^{1/6(+)}(VV/VV)^{(+)1/6}$

The inner core structure displays two (+) 1/6 magnetic charges, one at each end. A (+)1/3 electric charge is positioned outside the magnetic inner core, as shown in C; also appearing and functioning like a halo.

This unique type of charge and photonic structure, a(+)1/3 electric charge surrounding an inner core of twin magnetic charges of (+)1/6 at each of the ends, is herein defined as <u>a unit of electromagnetic force;</u> which in excess multiples produces (+) entities.

Z Boson

The W+ boson and W- boson are both formed in a magnetic phase dimension, while the Z boson is formed in the electric phase dimension. This will be shown in our discussion of the phases in Part III.

Once formed, the Z boson exhibits a zero electric charge, as well as simultaneously presenting opposite magnetic charges of (+) 1/6 and (-) 1/6, as follows.

E. $[^{(-1/6}(PP/VV)^{(+)1/6}]^{0}$

It's zero electric charge permits it to take part in electric coupling reactions with (+) 1/3 W+ bosons, (-) 1/3 W- bosons and with like Z bosons to produce ten types of three-part structures (triads).

Chart I

The Ten Triads

The W+, W- and Z bosons are components of ten different triads, which are divided into three different types of boson categories designated on p.17.

I. Containing only one type of boson (3):

1. Electron $[(W-)(W-)(W-)]^{(-)1}$
2. Positron $[(W+)(W+)(W+)]^{(+)1}$
3. Neutrino $[(Z)(Z)(Z)]^{(0)}$

II. Containing only two types of bosons (6):

1. Down quark $[(Z)(Z)(W-)]^{(-)1/3}$
2. Antidown quark $[(Z)(Z)(W+)]^{(+)1/3}$
3. Up quark $[(Z)(W+)(W+)]^{(+)2/3}$
4. Antiup quark $[(Z)(W-)(W-)]^{(-)2/3}$
5. Gluon $[(W-)(W-)(W+)]^{(-)1/3}$
6. Antigluon $[(W+)(W+)(W-)]^{(+)1/3}$

III. Containing all three types of bosons (1):

1. Antineutrino $[(W-)(Z)(W+)]^{(0)}$

These ten triads will be discussed at length below.

1. Electron (e-)

A. $[3(W-)]^{(-)1}$

B. $[(PP/PP)(PP/PP)(PP/PP)]^{(-)1}$

C. $[(P{=}P)\text{-}/\text{-}(P{=}P)^{(-)1/3}\text{-}/\text{-}(P{=}P)\text{-}/\text{-}(P{=}P)^{(-)1/3}\text{-}/\text{-}(P{=}P)\text{-}/\text{-}(P{=}P)^{(-)1/3}]^{(-)1}$

Of the three above versions of the electron's (e-) structure and composition, A and B are for quick appraisals and comparisons relative to other triads; whereas, C is an in-depth and exact depiction of all the photons and the bonds present, and is useful for intensive study. All of the ten triads will be presented in these three structural and compositional versions.

Using diagram 1.C. we note that there are three constituent W- bosons, $(P{=}P)\text{-}/\text{-}(P{=}P)^{(-)1/3}$, in the electron; each one depicted with its four P photons and two types of bonds, = and -/-, which weld the electron together.

The = bond is a (-) photon to (-) photon magnetic bond, joining the two P photons and producing a (-) monopole, $(P{=}P)^{(-)1/6}$.

The -/- bond is produced by two (-) monopoles bonding to

one another, creating the W- boson, (P=P)-/-(P=P)$^{(-)1/3}$. It is an electric (-) monopole to (-) monopole bond. There are three of these in 1C.

The -/- bond also appears between each of two W- bosons in an electron, fusing and bonding them together by a (-) monopole to (-) monopole bond, thus adding two additional -/- bonds in the electron to rivet the three W- bosons together into an electron, for a total of five

There are twelve P photons present in 1.C., each adding their (-)1/12 charge together for a total charge of (-) 1 for the electron.

One can also note the presence of six monopoles (P=P), each counted at (-)1/6, also a (-)1 total. Finally, each of the three W- bosons, [(P=P)-/-(P=P)], has a (-)1/3 charge, also adding up to (-)1 for the three.

2. <u>Positron (Antielectron) (e+)</u>

A. [3(W+)]$^{(+)1}$

B. [(VV/VV)(VV/VV)(VV/VV)]$^{(+)1}$

C. [(V$_+^+$V)+/+(V$_+^+$V)$^{(+)1/3}$+/+(V$_+^+$V)+/+(V$_+^+$V)$^{(+)1/3}$+/+(V$_+^+$V)+/+(V$_+^+$V)$^{(+)1/3}$]$^{(+)1}$

If one compares the above positron's [3(W+)]$^{(-)1}$ structure (2A) with the electron's [3(W-)]$^{(-)1}$ structure (1A), it is evident that they are total opposites on the electric scale. For this reason, the positron's other appellation, antielectron, may be more meaningful in summing up its essence.

The positron is composed of three W+ bosons, each of which is represented in 2.C. as (V$_+^+$V)+/+(V$_+^+$V)$^{(+)1/3}$. The bond between each of the V photons (V) therein is a magnetic (+) photon to (+) photon type: $_+^+$. The +/+ bond, on the other hand, fuses and bonds the two (+) monopoles together into a W+ boson. The +/+ bond is an electric (+) monopole to (+) monopole bond.

In turn, a (+) boson to (+) boson bond, +/+, links two W+ bosons together. The positron requires two of these bonds to link all three W+ bosons into a complete positron structure, which then displays a (+) 1 electric charge; a total of five +/+, as seen in 2C.

3. <u>Neutrino</u> (V_{E})
   A. $[3(Z)]^{0}$
   B. $[(PP/VV)(PP/VV)(PP/VV)]^{0}$
   C. $[(P=P)\text{-}/+(V_{+}{}^{+}V)^{0}+/\text{-}(P=P)\text{-}/+(V_{+}{}^{+}V)^{0}+/\text{-}(P=P)\text{-}/+(V_{+}{}^{+}V)^{0}]^{0}$
4. <u>Antineutrino</u> ($\overline{V}_{E}$)
   A. $[(W\text{-})(Z^{0})(W+)]^{0}$
   B. $[(PP/PP)(PP/VV)(VV/VV)]^{0}$
   C. $[(P=P)\text{-}/\text{-}(P=P)^{(-)1/3}\text{-}/+(P=P)\text{-}/+(V_{+}{}^{+}V)^{0}+/+(V_{+}{}^{+}V)+/+(V_{+}{}^{+}V)^{(+)1/3}]^{0}$

We shall discuss the structure and composition of the above two triads, 3 and 4, together, as they both have a net zero electric charge and, in many ways, parallel each other.

The two triads, however, are very dissimilar in structure: $[3(Z)]^{0}$ versus $[(W\text{-})(Z^{0})(W+)]^{0}$.

The presence of a zero electric charge, however, permits both of them to slide past many other reactive structures without being engaged in a reaction process.

Both types of neutrinos, however, cannot be constituents of a proton, neutron, antiproton or antineutron; but can be formed from each of their photon constituents by rearrangement, due to the impingement and reaction with another reactive body, such as a positron or electron. Each is then always ejected at the end of such a reaction. This is due entirely to the presence of their zero electrical charge. However, both types of neutrinos can react with protons, neutrons, antiprotons and antineutrons due to their inner magnetic charges, such as are shown in 3.C. and 4.C.

Their end magnetic (+) or (-) 1/6 charges act like hooks, and are the cause of a small number of reactions both can undergo.

Neutrinos and antineutrinos are formed as reaction products in respective proton and neutron decay processes; as well as in proton and electron interactions.

In these cases, neutrinos and antineutrinos are formed in the breakdown of the quark and the gluon components of the proton and neutron.

The remaining six triads (up and down quarks, antiup and antidown quarks, gluons and antigluons) have either a (+) or a (-) magnetic charge at both ends of their internal structure; or one of each type at each end. Thus, a wide variety of different spatially structured bodies can be created, each of which can be given a specific "color" description. We will describe the different "colors" for each of these triads at the time of their respective presentation.

Only a triad that has both "color" and displays a (+) or (-) unit of electric charge can become a component structure of a proton, antiproton, neutron or antineutron. Thus, only the following six triads are eligible: down, up, antidown and antiup quarks; gluon and antigluons (the latter two in the form of a gluon/antigluon meson).

The electron and the positron have no "color" possible due to their homogenous all P or V content, respectively. The neutrino and antineutrino display a net zero unit of electric charge. Therefore, all four are thus precluded from being a structural component of a proton[(+)1], neutron or their antiparticles; although each of the four can be formed as by-product in various reactions these four enter into; or in a structural re-arrangement process, such as a proton or neutron decay process; but each of these four is then ejected from the reaction area at the conclusion.

Magnetic and Electrical Bonds

Before we describe each of these remaining six triads, an understanding of the difference in the nature of the magnetic and electrical bonds will be useful.

There are four entities we define as magnetic: a P photon and a V photon, each respectively displaying a (-) 1/12 or a (+) 1/12 charge; a (+) monopole (VV) and a (-) monopole (PP), each respectively exhibiting a (+) 1/6 or (-) 1/6 charge.

The electrical entities are as follows: the three bosons: W+ with a (+) 1/3 charge; W- with its (-) 1/3 charge; and Z which also carries an electrical charge, a neutral one referred to as zero; the up quark at (+) 2/3; the antiup quark at (-) 2/3; the down quark and gluon both at (-)1/3; the antidown quark and antigluon both at (+) 1/3; the neutrino, antineutrino, neutron and antineutron, all carrying a neutral electric, or zero, charge; the positron and proton at (+) 1 and the electron and antiproton at (-) 1. All of the above charges, whether defined as magnetic or electrical, are fractional except those at (+) or (-) 1, or zero.

If, however, we designated the P and V photons as having a (-) 1 and a (+) 1 charge, respectively, instead of (-) 1/12 and (+) 1/12, then the (+) and (-) monopoles would be: (+) 2 and (-) 2; the W+ and W- boson at (+) 4 and (-) 4; the down quark and the gluon at (-) 4; the antidown quark and the antigluon at (+) 4; the up quark and the antiup quark at (+) 8 and (-) 8, respectively; the positron at (+) 12 and the electron and antiproton at (-) 12. The neutrino, antineutrino, neutron and antineutron still remain at zero.

As one can see from this reordering of charges, there are no fractional charges, only whole number, which indicates order of magnitude differences.

However, we will continue, herein, to use the original historical fractional and whole number designations due to the possible greater confusion if we substitute the 1-12 system.

One reason for presenting and illustrating this 1-12 system is to acquaint the reader with the concept of whole number magnitude differences to describe what really is occurring, which the fractional and whole number system seems to obfuscate.

The reason for differentiating (+) and (-) 1/12 and (+) and (-) 1/6 charges as magnetic, and all others, starting with (+) and (-) 1/3 and including all others at higher (+) and (-) values, as electrical, is to clearly indicate that the former are part of a separate and distinctly different phase, the magnetic; while the latter's electric structure originated in a different phase, the electric, which will be discussed in greater detail later.

5. <u>Down Quark</u> (d)

A. $[2\ (Z)(W-)]^{(-)1/3}$

B. $[(PP/VV)(VV/PP)(PP/PP)]^{(-)1/3}$

C. $[^{(-)1/6}(P{=}P)-/+(V_+^+V)+/+(V_+^+V)+/-(P{=}P)-/-(P{=}P)-/-(P{=}P)^{(-)1/6}]^{(-)1/3}$

The down quark is the first of six fractionally charged triads which completes the list of ten triads.

The down quark and the up quark can only become structural components of the proton and neutron, while the antidown quark and the antiup quark can only become components of the antiproton and antineutron. This will be shown, in a later chapter, to be due to an extraordinary system which is in place that only permits these combinations to form.

Blue Color: (P=P)/ /(P=P)

In 5.C. P.21, we note that both end monopoles are (P=P); thus, both ends also show a (-) 1/6 magnetic charge.

We will now arbitrarily label a down quark, antidown quark, up quark, antiup quark, gluon or antigluon that has both of its two end monopoles as (P=P) to be a "blue" color; or, in this specific case (5.C.), as a blue down quark.

Yellow Color: (VV)/ / (P=P)

If, however, the left end boson in 5.C., ${}_{(-)1/6}(P=P)$ -/+ (V_+^+V), were reversed 180° to become ${}_{(+)1/6}(V_+^+V)$+/- (P=P) during the formation process of this down quark, then the structure of the resultant down quark would be as follows:

5 D. $[{}_{(+)1/6}(V_+^+V)+/-(P=P)/(V_+^+V)+/-(P=P)-/-(P=P)-/-(P=P)_{(-)1/6}]_{(-)1/3}$

In 5.D., above, we now have a (V_+^+V) at the left end position, thus producing a (+) 1/6; and a (P=P) with a (-) 1/6 charge at the right end of the inner core structure of the down quark. If two such different charges also appeared on at the end positions of any inner core of any of the other five triad structures (up quark, antidown quark, etc.), we would arbitrarily refer to such a triad as a "yellow" color triad; or as a yellow down quark in this specific case (5.D.).

Red Color: (VV)/ /(VV)

One last rearrangement of the down quark structure can be made, which is below:

5 E. $[^{(+)1/6}(V_{+}{}^{+}V)+/-(P{=}P)-/-(P{=}P)-/-(P{=}P)-/-(P{=}P)-/+(V_{+}{}^{+}V)^{(+)1/6}]^{(-)1/3}$

Here, the two end monopoles in 5.E. are both $(V_{+}{}^{+}V)$, which are (+) 1/6 structures. This same type of magnetic charge on both end monopoles of the inner core is herein arbitrarily called a "red" color; or, in this instance, a red down quark. All of the other five triads having a (+) 1/6 charged $(V_{+}{}^{+}V)$ at two end positions of their inner core will also be labeled a red up quark, red antidown quark, etc.

In Chart II, following, we will see that there is only one red down quark possible: dR1 (down red 1).

However, there are also six yellow down quark variants, as well as five blue down quark variants.

These twelve possible structural variants of a down quark, listed in Chart II, are derived from the original composition of two Z bosons (VV/PP) and a single W- boson (PP/PP), depicted in 5.A. and 5.B, p.21.

If we examine the six yellow down quarks, we note that dY1, dY2 and dY3 are each followed in order by dY1r, dY2r and dY3r, respectively. The small "r" designates the structure as the 180° reverse of the same designated down quark without the r. So, dY1r is dY1 turned around 180° spatially. The same is true with dY2r and dY2, as well as dY3r and dY3.

All r-types are the same or equivalent to their reverse non-r-types and can be substituted for them and vice-versa.

The reason that under the heading "Structure" we indicate which end is the "bottom" or "top" is to allow correct diagrams to be created. When a coupling reaction takes place with other down quarks or up quarks to form a proton or a neutron, both of which have triangle-type structures, the "bottom" boson of a down quark must be located at the bottom and the "top" boson must be located at the top. This is exactly what is shown in diagram 5.F., p.25,

Chart II

Down Quarks (d): Red, Yellow and Blue Types

COLOR	TYPE	CHARGE	STRUCTURE Bottom		Top
1. Red	dR1	(-) 1/3	+	(VV/PP)(PP/PP)(PP/VV)	+
1. Yellow	dY1	(-) 1/3	-	(PP/PP)(PP/VV)(PP/VV)	+
2. Yellow	dY1r	(-) 1/3	+	(VV/PP)(VV/PP)(PP/PP)	-
3. Yellow	dY2	(-) 1/3	-	(PP/PP)(VV/PP)(PP/VV)	+
4. Yellow	dY2r	(-) 1/3	+	(VV/PP)(PP/VV)(PP/PP)	-
5. Yellow	dY3	(-) 1/3	-	(PP/VV)(PP/PP)(PP/VV)	+
6. Yellow	dY3r	(-) 1/3	+	(VV/PP)(PP/PP)(VV/PP)	-
1. Blue	dB1	(-) 1/3	-	(PP/VV)(VV/PP)(PP/PP)	-
2. Blue	dB1r	(-) 1/3	-	(PP/PP)(PP/VV)(VV/PP)	-
3. Blue	dB2	(-) 1/3	-	(PP/VV)(PP/VV)(PP/PP)	-
4. Blue	dB2r	(-) 1/3	-	(PP/PP)(VV/PP)(VV/PP)	-
5. Blue	dB3	(-) 1/3	-	(PP/PP)(VV/PP)(VV/PP)	-

which is a neutron.

If, for example, the position of the down quark, dY1r, below, was changed by transfer to be in the base position in 5.F, where the up quark is, and the up quark was placed in the dY1r position, we would then have the diagram 5.G, p.26.

Thus, in 5.G., the down quark, dY1r, positioned between the down quark, dY1, and the up quark, requires that the down quark dY1r "bottom" position to be on the left side and the "top" on the right side, exactly as it appears in Chart II.

Finally, the twelve different down quark variant "colors" shown in Chart II would be termed isomers by chemists. They each modify the properties of the structure of which they become a part; in the same way each isomer in a chemical reaction produces a product that has different properties than a product based on

continued on P.27

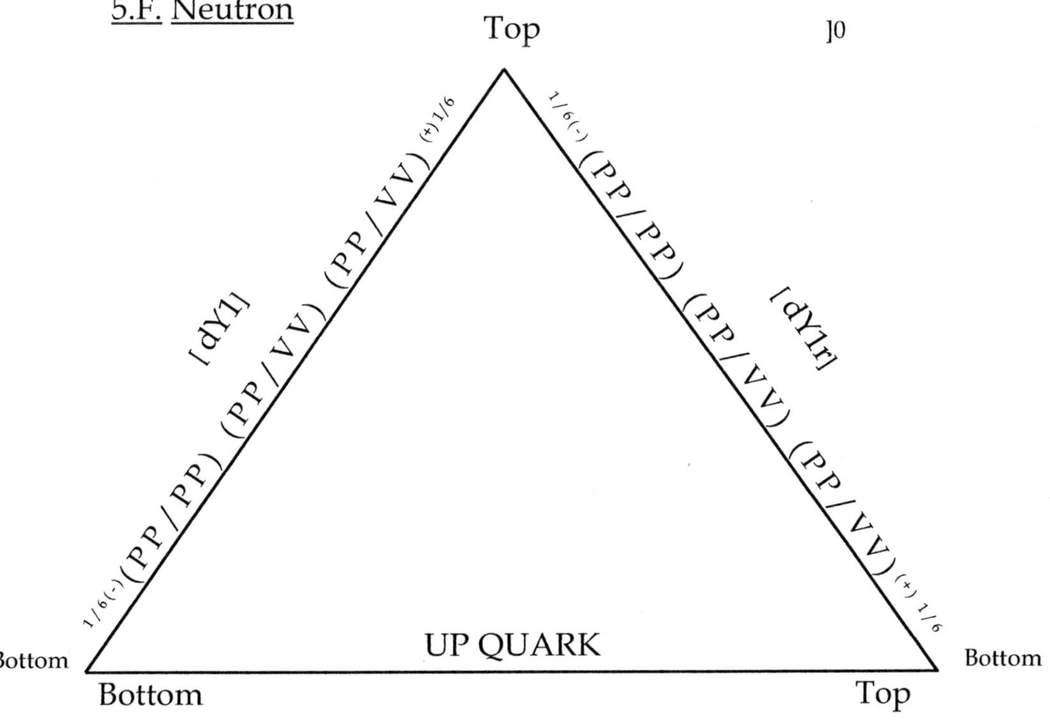

5.G. Neutron

Top
]0
[dY1]
1/6(-)(PP/PP) (PP/VV) (PP/VV) (+)1/6
UP QUARK
Bottom
Bottom
Bottom
(+)1/6(VV/PP) (VV/PP) (PP/PP)(-)1/6
[dY1r]

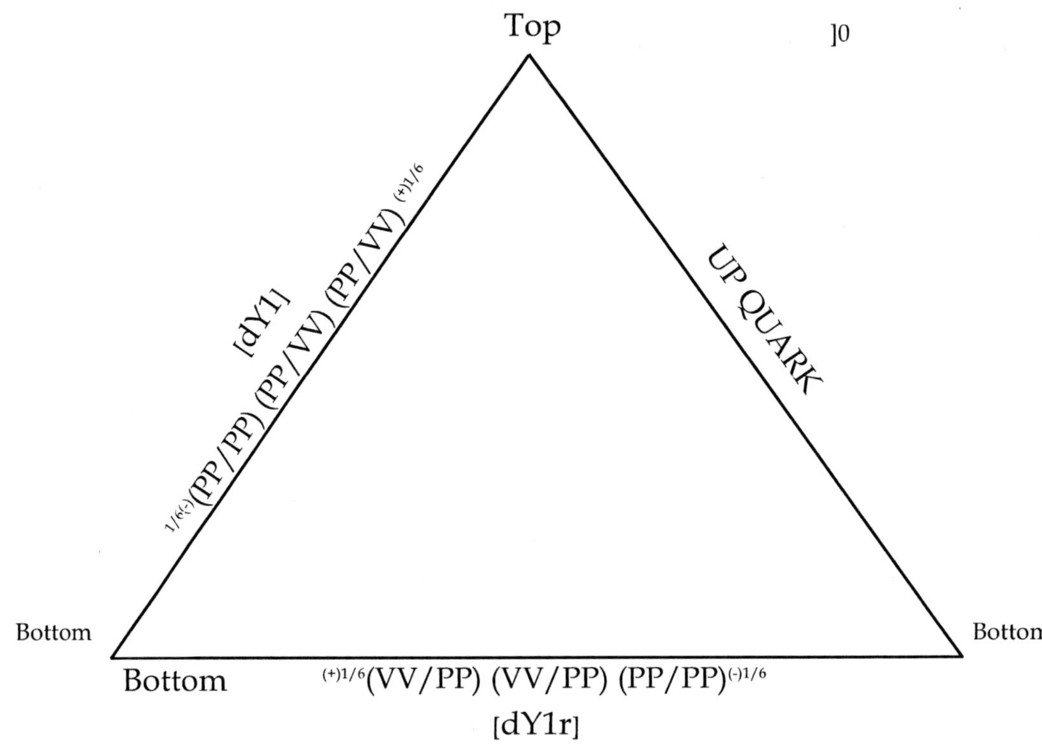

another isomer.

6. <u>Antidown Quark ($\bar{d}$)</u>

A. $[2(Z^0)\ (W+)]^{(+)1/3}$

B. $[(PP/VV)(VV/PP)(VV/VV)]^{(+)1/3}$

C. $[^{(-)1/6}(P{=}P)\text{-}/{+}(V_+^+V){+}/{+}(V_+^+V){+}/\text{-}(P{=}P)\text{-}/{+}(V_+^+V){+}/{+}(V_+^+V)^{(+)1/6}]^{(+)1/3}$

The antidown quark ($\bar{d}$) has a net electric charge of (+) 1/3, in contrast to the down quark's (d) net (-) 1/3 electric charge.

Also, the down quark's W- boson is replaced in the antidown quark by a W+ boson. Both types of down quarks each contain two Z^0 bosons.

Like the down quark, the antidown quark has twelve variant structures, also in three colors: red, yellow and blue. These are shown in Chart III, which follows on p.28.

In Chart III, the blue antidown quark, $\bar{d}$ B1, is the sole blue structure. The red has three types: $\bar{d}$ R1, $\bar{d}$ R2 and $\bar{d}$ R3; plus two reverse (180°) orientation "r" types: $\bar{d}$R1r and $\bar{d}$R2r; while yellow also has three types: $\bar{d}$ Y1, $\bar{d}$ Y2 and $\bar{d}$ Y3; plus three of the reverse (180°) orientation types: $\bar{d}$ Y1r, $\bar{d}$ Y2r and $\bar{d}$ Y3r, a total of twelve types in three "colors".

Although we have herein identified and depicted twelve down quarks and twelve antidown quarks, in reality there are only seven true down quark and seven true antidown quark structures. Five of the twelve in each of these two quark categories is a 180^0 reverse (r) of five of the respective quarks. However, each of the five reverse structures in each category still must be seen as a seperate quark structure in that it has an effect on the final completed structure it is a part of.

As a parallel, each of the six up quarks and each of the six antiup quarks has 3 of their structures as 180^0 reverse; while one of the three gluons and one of the three antigluons is also a 180^0 reverse structure.

Chart III

Antidown Quarks ($\bar{d}$) : Red, Yellow and Blue Types

COLOR	TYPE	CHARGE	STRUCTURE Bottom		Top
1. Red	$\bar{d}$ R1	(-) 1/3	+	(VV/VV)(VV/PP)(PP/VV)	+
2. Red	$\bar{d}$ R1r	(-) 1/3	+	(VV/PP)(PP/VV)(VV/VV)	+
3. Red	$\bar{d}$ R2	(-) 1/3	+	(VV/VV)(PP/VV)(PP/VV)	+
4. Red	$\bar{d}$ R2r	(-) 1/3	+	(VV/PP)(VV/PP)(VV/VV)	+
5. Red	$\bar{d}$ R3	(-) 1/3	+	(VV/PP)(VV/VV)(PP/VV)	+
1. Yellow	$\bar{d}$ Y1	(+) 1/3	+	(VV/VV)(VV/PP)(VV/PP)	-
2. Yellow	$\bar{d}$ Y1r	(+) 1/3	-	(PP/VV)(PP/VV)(VV/VV)	+
3. Yellow	$\bar{d}$ Y2	(+) 1/3	+	(VV/VV)(PP/VV)(VV/PP)	-
4. Yellow	$\bar{d}$ Y2r	(+) 1/3	-	(PP/VV)(VV/PP)(VV/VV)	+
5. Yellow	$\bar{d}$ Y3	(+) 1/3	+	(VV/PP)(VV/VV)(VV/PP)	-
6. Yellow	$\bar{d}$ Y3r	(+) 1/3	-	(PP/VV)(VV/VV)(PP/VV)	+
1. Blue	$\bar{d}$ B1	(+) 1/3	-	(PP/VV)(VV/VV)(VV/PP)	-

Each reverse structure is identified by an (r) at the end of its designation: UR1r, UY1r, gY1r, āY1r, etc.

In coupling reactions in which the antidown quark becomes a component of an antiproton or antineutron, their final triangular structures follow the same rules relating to "top" and "bottom" positions as just illustrated in the down quark section.

All the antidown quarks, as well as antiup quarks, however, have disappeared from the existent universe, transformed into down and up quarks, respectively, due entirely to their being components of antiprotons and antineutrons when these structures were transformed by reactions with positrons into becoming protons; or electrons, neutrinos, antineutrinos and an additional positron. This will be detailed later in the antiproton and antineutron sections.

7. <u>Up Quarks (U)</u>

   A. $[(Z)\ 2(W+)]^{(+)2/3}$

   B. $[(PP/VV)(VV/VV)(VV/VV)]^{(+)2/3}$

   C. $[^{(-)1/6}(P{=}P)\text{-}/\text{+}(V_{+}^{+}V)\text{+}/\text{+}(V_{+}^{+}V)\text{+}/\text{+}(V_{+}^{+}V)\text{+}/\text{+}(V_{+}^{+}V)\text{+}/\text{+}(V_{+}^{+}V)^{(+)1/6}]^{(+)2/3}$

The up quark's composition, as seen in 7.A., above, is one Z boson and two W+ bosons, thereby displaying a (+) 2/3 electric charge.

In 7.C. above, we note the presence of a (-) 1/6 at the left end of its inner core structure and a (+) 1/6 at the other end. By definition, then, the up quark depicted here is yellow, UY1, (up yellow 1).

Looking at Chart IV, following, we see that it and its 180° reverse structure, UY1r, are, in fact, the only two yellow up quarks. There are also four red up quarks: UR1 and UR2 and their two respective 180° reverse structures, UR1r and UR2r.

No blue up quark can be constructed and are, therefore, absent from Chart IV. Thus, a total of six color up quark variants.

Chart IV

Up Quarks (U): Red and Yellow Types

COLOR	TYPE	CHARGE	Bottom	STRUCTURE	Top
1. Red	UR1	(+) 2/3	+	(VV/PP)(VV/VV)(VV/VV)	+
2. Red	UR1r	(+) 2/3	+	(VV/VV)(VV/VV)(PP/VV)	+
3. Red	UR2	(+) 2/3	+	(VV/VV)(PP/VV)(VV/VV)	+
4. Red	UR2r	(+) 2/3	+	(VV/VV)(VV/PP)(VV/VV)	+
5. Yellow	UY1	(+) 2/3	-	(PP/VV)(VV/VV)(VV/VV)	+
6. Yellow	UY1r	(+) 2/3	+	(VV/VV)(VV/VV)(VV/PP)	-

The up quark is found as a component of the proton or neutron. It is not stable in an antiproton or antineutron, as will be shown in a subsequent chapter.

8. <u>Antiup Quark</u> ($\overline{U}$)

A. $[(Z)\ 2(W-)]^{(-)2/3}$

B. $[(PP/VV)(PP/PP)(PP/PP)]^{(-)2/3}$

C. $[^{(-)1/6}(P{=}P)$ -/+ $(V_{+}{}^{+}V)$ +/- $(P{=}P)$ -/- $(P{=}P)$ -/- $(P{=}P)$ -/- $(P{=}P)^{(-)1/6}]^{(-)2/3}$

The constituents of the antiup quark displayed in 8.A. are a single Z boson and 2 W- bosons, projecting a (-) 2/3 electric charge.

Like the up quark, the antiup quark has a total of six color variants. Chart V, p.31, lists and describes all of them: four blue types: $\overline{U}$B1 and $\overline{U}$B2 and their respective 180° reverse structures, $\overline{U}$B1r and $\overline{U}$B2r; and two yellow structures: $\overline{U}$Y1 and its 180° reverse $\overline{U}$Y1r.

There are no red color units possible.

Chart V

Antiup Quarks ($\overline{U}$): Blue and Yellow Types

COLOR	TYPE	CHARGE	STRUCTURE Bottom Top
1. Blue	$\overline{U}$B1	(-) 2/3	- (PP/PP)(PP/PP)(VV/PP) -
2. Blue	$\overline{U}$B1r	(-) 2/3	- (PP/VV)(PP/PP)(PP/PP) -
3. Blue	$\overline{U}$B2	(-) 2/3	- (PP/PP)(PP/VV)(PP/PP) -
4. Blue	$\overline{U}$B2r	(-) 2/3	- (PP/PP)(VV/PP)(PP/PP) -
5. Yellow	$\overline{U}$Y1	(-) 2/3	- (PP/PP)(PP/PP)(PP/VV) +
6. Yellow	$\overline{U}$Y1r	(-) 2/3	+ (VV/PP)(PP/PP)(PP/PP) -

9. Gluon (g)
A. [2(W-) (W+)]$^{(-)1/3}$
B. [(PP/PP)(PP/PP)(VV/VV)]$^{(-)1/3}$
C. [$^{(-)1/6}$(P=P)-/-(P=P)-/-(P=P)-/-(P=P)-/+($V_{+}{}^{+}V$)+/+($V_{+}{}^{+}V$)$^{(+)1/6}$]$^{(-)1/3}$

10. Antigluon ($\bar{a}$)
A. [2(W+) (W-)]$^{(+)1/3}$
B. [(VV/VV)(VV/VV)(PP/PP)]$^{(+)1/3}$
C. [$^{(+)1/6}$($V_{+}{}^{+}V$)+/+($V_{+}{}^{+}V$)+/+($V_{+}{}^{+}V$)+/+($V_{+}{}^{+}V$)+/-(P=P)-/-(P=P)$^{(-)1/6}$]$^{(+)1/3}$

The gluon and its antiparticle, the antigluon, pictured above, are structurally quite different from any of the preceding four triads in that they both are absent a Z boson, thus having only W+ and W- bosons present in their structure.

Chart VI.1

Gluon (g): Yellow and Blue Types

COLOR	TYPE	CHARGE	STRUCTURE Bottom … Top
1. Yellow	gY1	(-) 1/3	- (PP/PP)(PP/PP)(VV/VV) +
2. Yellow	gY1r	(-) 1/3	+ (VV/VV)(PP/PP)(PP/PP) -
3. Blue	gB1	(-) 1/3	- (PP/PP)(VV/VV)(PP/PP) -

Chart VI.2

Antigluon (a): Yellow and Red Types

COLOR	TYPE	CHARGE	STRUCTURE Bottom … Top
1. Yellow	āY1	(+) 1/3	+ (VV/VV)(VV/VV)(PP/PP) -
2. Yellow	āY1r	(+) 1/3	- (PP/PP)(VV/VV)(VV/VV) +
3. Red	āR1	(+) 1/3	+ (VV/VV)(PP/PP)(VV/VV) +

Synthesis of Gluon/Antigluon Meson (g/ā)

We have just described the last two of the ten independent triads: the gluon and the antigluon, each having two "colors", respectively. The gluon has two yellows and one blue, while the antigluon has two yellows and one red.

The synthesis of a gluon/antigluon meson (g/ā) is graphically depicted in Chart VII,p.33, where 1., an electron, and 2., a positron, interact with each other, resulting in a V-shaped structure, which consists of a gluon and an antigluon bonded together into a gluon/antigluon (ā Y1/ ā R1) meson, which is seen in 3., p.33. Here, the left arm is a gluon, a yellow gY1 type; while the right arm is the antigluon, a red ā R1 type.

One could argue that the aforesaid meson is really a combined electron and positron, with each of the two exchanging a respective boson: (PP/PP) for a (VV/VV).

Chart VII

Gluon/Antigluon Meson Synthesis

1. [(PP/PP)(PP/PP)(PP/PP)]$^{(-)1}$

<u>Electron</u>

+

2. [(VV/VV)(VV/VV)(VV/VV)]$^{(+)1}$

<u>Positron</u>

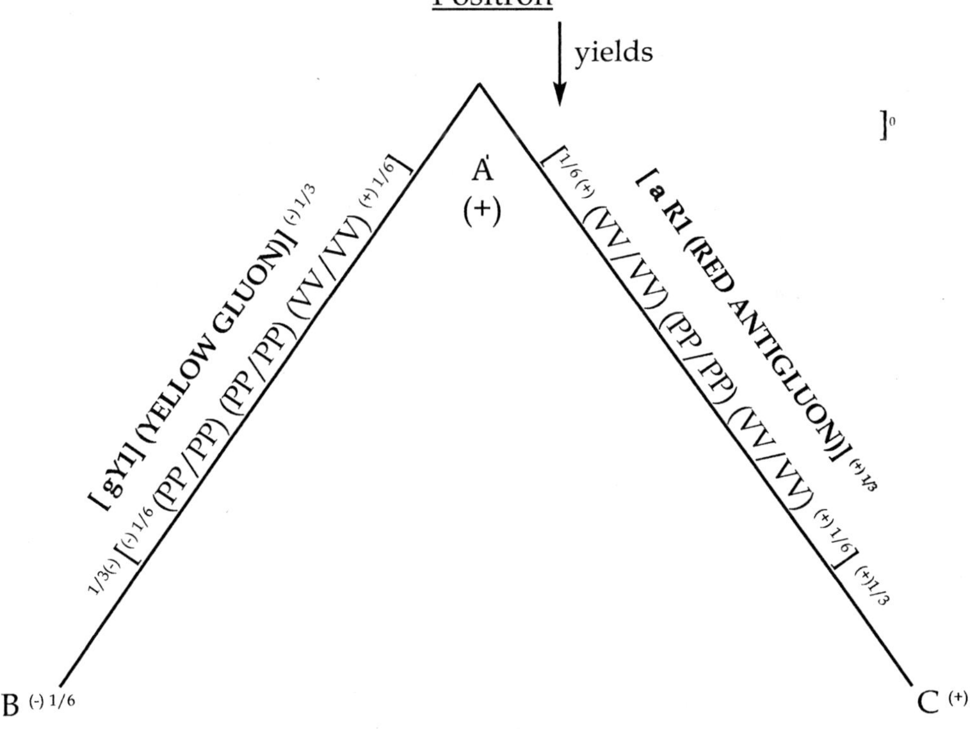

3. <u>[Gluon/Antigluon (gY1/ ā R1) Meson]^{0} (S3 Meson)</u>

The gluon and the antigluon components of this type of meson are never found as independent, separate entities. They always must be bonded together as a meson. We will, however, treat them as independent structural units in the meson for clarity of discussion.

The yellow gY1 gluon in Chart VII.3, above, would carry a (-) 1/3 charge, and the red ā R1 antigluon would carry a (+) 1/3 charge, if they were able to be isolated. The net charge of the two combined as a meson is thus zero (0). This type of meson is identified as an S3, stable meson, in Chart VIII, p.37.

In chart VII.3, p.33, a (+) to (+) magnetic bond between the end gY1's/VV(+) and the end ā Y1r's/VV(+), at a point shown as Á, is produced, which, in turn, is represented by a (+) under the Á.

In Chart VIII, S2, p.36, the magnetic bond produced at Á is a (-) to (-) type, and represented by a (-) under the Á. In the above two examples, and in Chart VIII, S1 and S4, a bond between like charges at point Á is the key to stability of the meson; which stability is indicated by an "S" before 1, 2, 3 and 4, (p36 & 37.)

In contradistinction to these four stable g/ā mesons, S1 through S4, are the five unstable mesons, shown in Chart VIII, U1 through U5, (p.38-40). Each of these has a (+) to (-) bond at point Á, which is represented by a (0) under Á. The (0) indicates that a magnetic bond has not formed between the gluon and antigluon at point Á. The unstable meson is, however, held together weakly by an electric type (-) 1/3 (gluon) to (+) 1/3 (antigluon) bond. The "U" before 1, 2, 3, 4 and 5 indicates the potential unstable nature of the meson.

Each of the nine types of g/ā mesons was produced by a reaction between an electron and a positron in an electric dimension (E4). However, the heat of reaction, added to the high temperature already in place in E4, was so great as to elevate the resultant temperature of the gluon and antigluon meson to over 10 million K., thus transforming the meson and the immediate space surrounding the meson from an E4, an electric dimension, into an M10, a magnetic dimension.

In the M10 dimension, like attracts and bonds to like, while opposite charges repel: (+) to (+) and (-) to (-) bonds can thus be formed, while (+) to (-) charged entities repel each other, with no bond forming. It is only when these unbonded mesons are cooled to below 10 million K. that a (+) to (-) electric bond can then form between the (-) 1/3 gluon and the (+) 1/3 antigluon in what now has once more become the electric E4 dimension.

In an electric dimension, the magnetic bonds are very stable, since only at 10 million K. plus can these bonds be unraveled; while the electric (+) to (-) bond can be undone at any temperature below 10 million K. by a weak interaction.

If, instead of the red āR1 antigluon in S3 on p.37, in Chart VIII, a yellow āY1 antigluon is formed in its place, producing the g/ā meson shown in Chart VIII, U2, p.38, a (+) 1/6 to (-) 1/6 bond results at its midpoint (Á) and thus no magnetic bond would form (0) in a magnetic dimension (M10). The resultant meson could only be held together by an electric (-) 1/3 to (+) 1/3 gluon to antigluon bond when it reaches a temperature below 10 million K. in what then would be an electric dimension (E4).

All mesons having these opposite charge (+) to (-) bonds at the juncture point Á of the gluon and the antigluon in an electric dimension are unstable and subject to destruction by passing particles, energy fluxes and particle interaction. A (+) to (+) or (-) to (-) magnetic bond, however, at the same juncture point Á in an electric dimension is not affected by these actions and, thus, is stable.

Chart VIII lists and describes graphically four stable gluon/antigluon mesons (S1-S4) and five unstable ones (U1-U5).

Thus, in calculating the net charge on either a gluon or antigluon in any of the above mesons, a (PP/PP) is counted as (-)1/3, while the (VV/VV) is (+)1/3. Therefore, the gYI gluon in chart VIII, S1 and S3, is two (PP/PP) at (-)1/3 each, while the one (VV/VV) is (+) 1/3 ; producing a net charge on the gluon of (-)1/3.

Stable mesons, therefore, are held together by two types of bonds: a (+) to (+), or (-) to (-) magnetic bond at their Á juncture; as well as by an electric (-) 1/3 to (+) 1/3 bond, respectively produced by it's component gluon and antigluon in a subsequently colder elecric dimension.

Chart VIII
Gluon/Antigluon Mesons (g/ā)

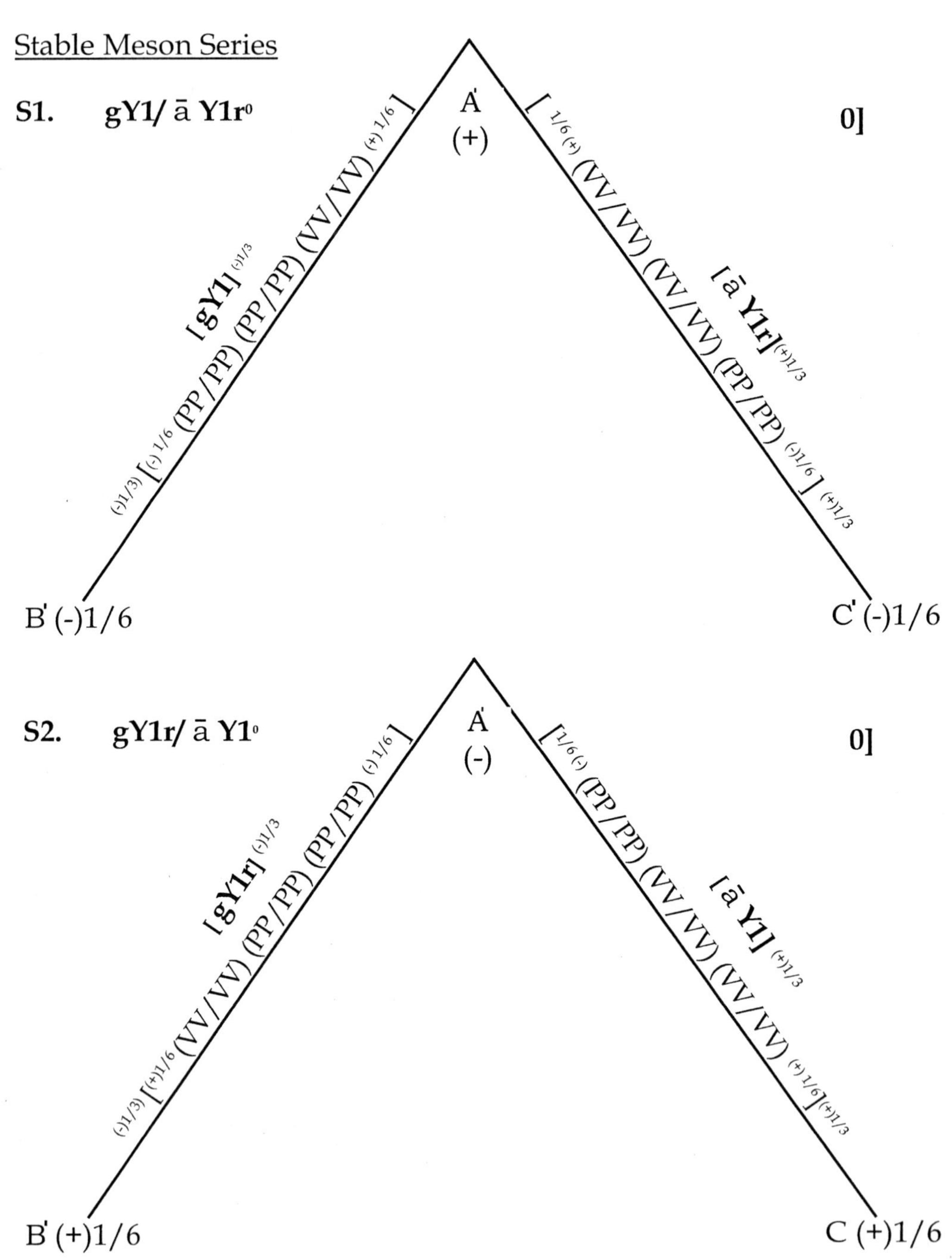

Chart VIII (continued)

Stable Meson Series (cont'd)

S3. gY1r/ ā R1⁰

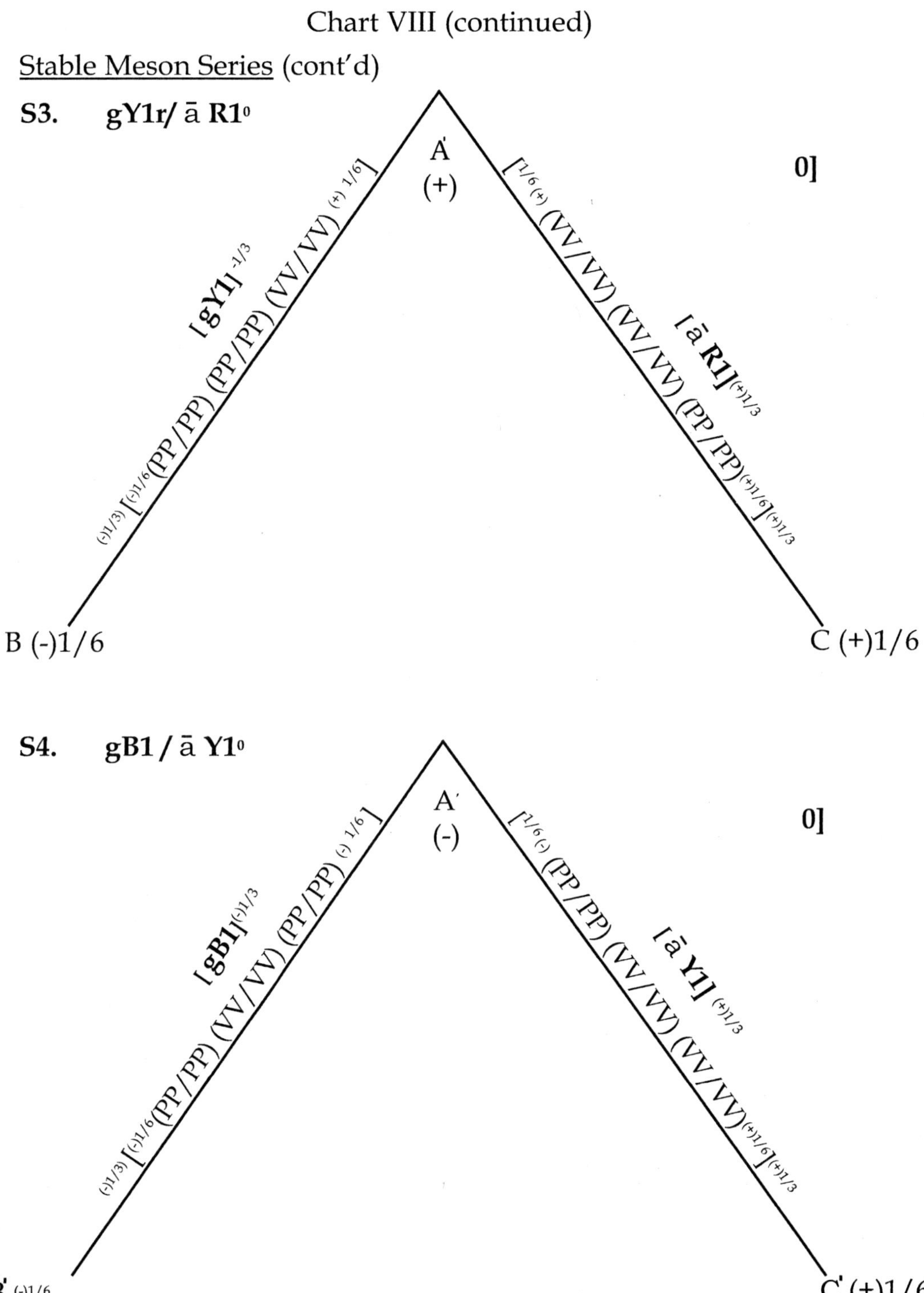

Chart VIII (continued)

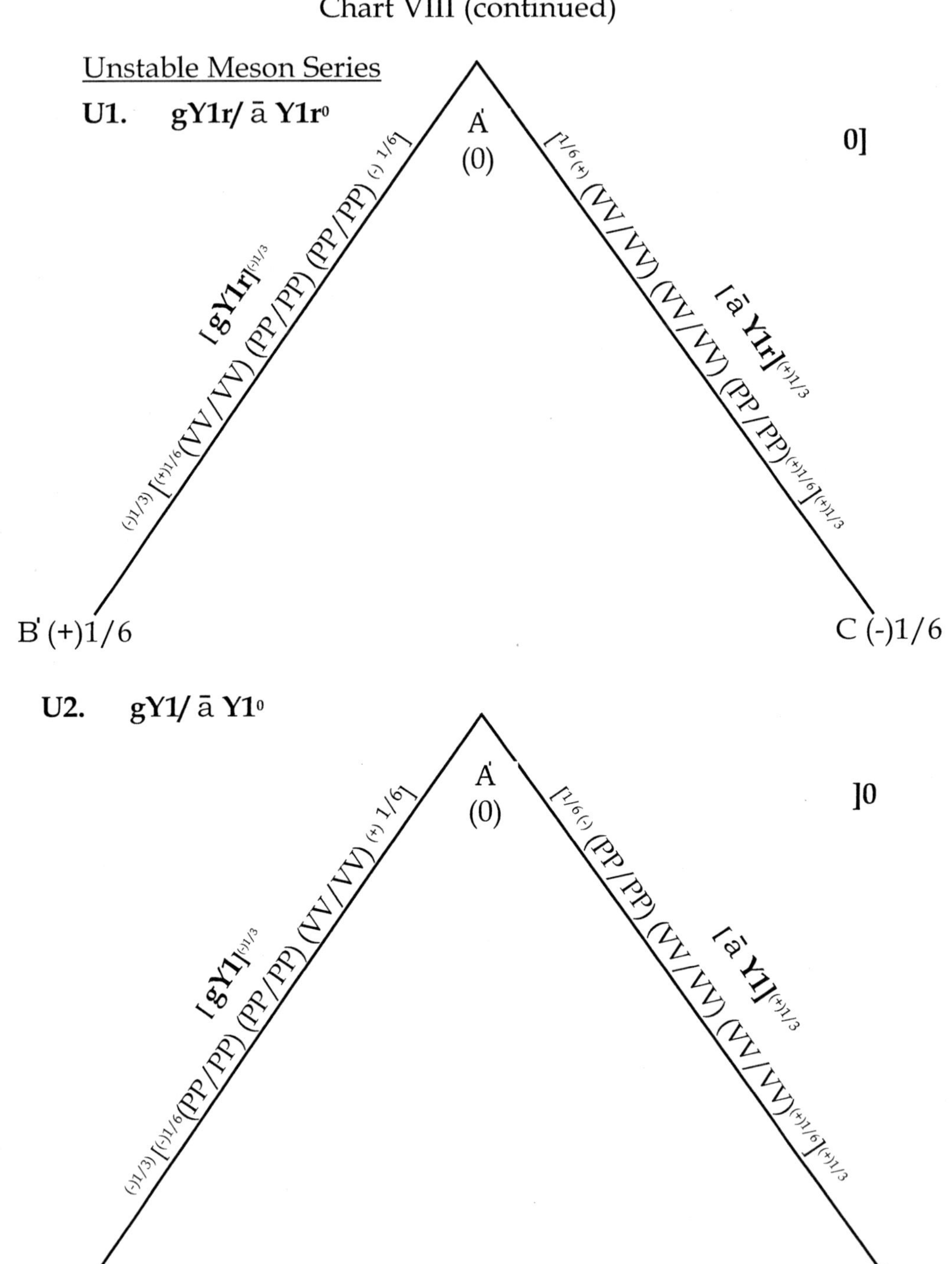

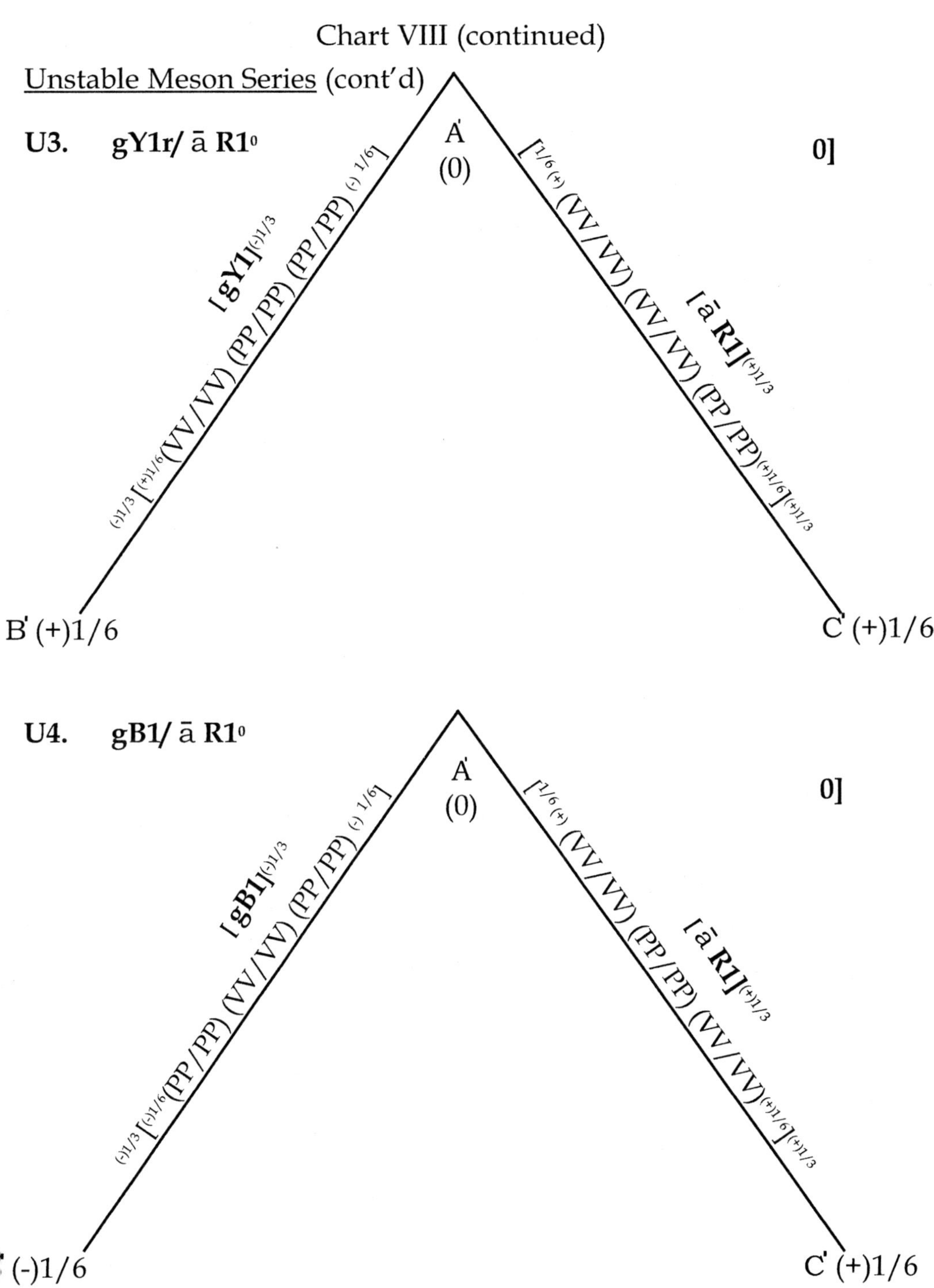
Chart VIII (continued)
Unstable Meson Series (cont'd)
U3. gY1r/ ā R1⁰
0]
A'
(0)
(-)1/3 [(+)1/6 (VV/VV) (PP/PP) (PP/PP) (-) 1/6]
[gY1](-)1/3
[1/6 (+) (VV/VV) (VV/VV) (PP/PP) (+)1/6](+)1/3
[ā R1](+)1/3
B' (+)1/6
C' (+)1/6
U4. gB1/ ā R1⁰
0]
A'
(0)
(-)1/3 [(-)1/6 (PP/PP) (VV/VV) (PP/PP) (-) 1/6]
[gB1](-)1/3
[1/6 (+) (VV/VV) (PP/PP) (VV/VV) (+)1/6](+)1/3
[ā R1](+)1/3
(-)1/6
C' (+)1/6

Chart VIII (continued)

Unstable Meson Series (cont'd)

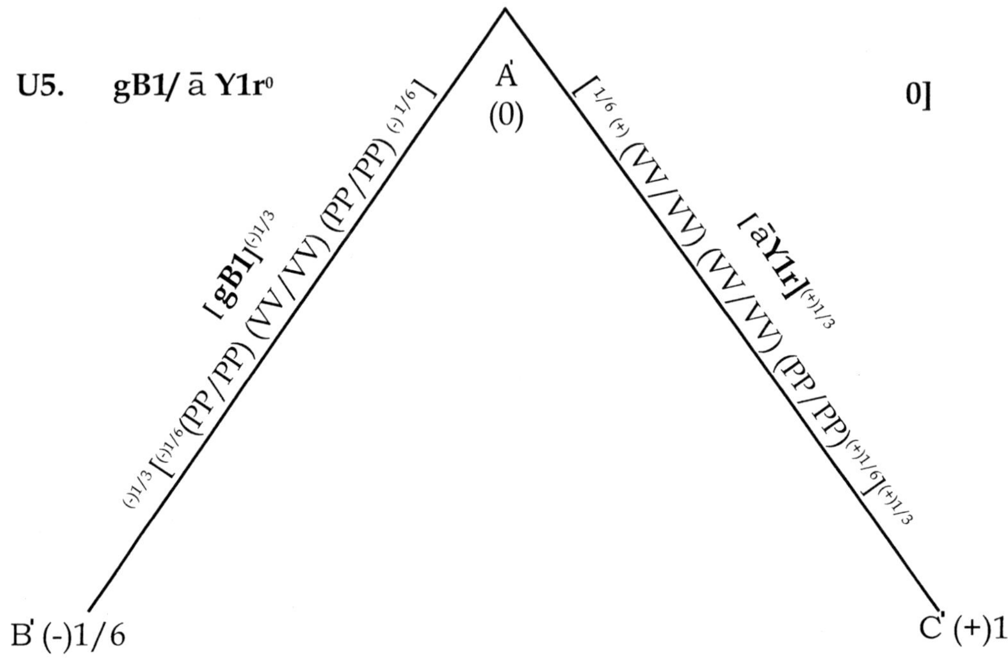

Unstable mesons can only rely on the latter type of bonding to hold them together, as their magnetic type bonds could not form.

Stable Mesons

In Chart VIII, p.36 & 37, we list and graphically portray four stable meson structures: S1, S2, S3 and S4. Each possesses either a (+) to (+) or a (-) to (-) connection at the midpoint nexus of the meson at point (Á).

Since all the bonds shown in the stable mesons are either (+) to (+) or (-) to (-) at Á, that area of the meson can be considered as a magnetic phase domain. All of the four stable mesons have like charge bonds at this juncture and, since like attract like in magnetic phase domains, the bond at the midpoint A' is a stable bond which was created in a magnetic phase reaction (above 10 million K.).

The two charges at each of the other two ends of the meson, B' and C', can only very slightly add or subtract from stability when they bond or fail to bond with adjacent charges on entities such as quarks and antiquarks.

Unstable Mesons

The five unstable mesons listed in Chart VIII, U1-U5, p.38-40, all display opposite charge bonds (+ to -) at the midpoint A'. Because opposites repel in magnetic phase domains, this type of bond does not form therein. The meson, however, is kept together in an electric phase domain, where opposites attract therein, by the (-) 1/3 gluon to (+) 1/3 antigluon electric bond, which can be easily ruptured; whereas, the magnetic type bond at point A', either a (+) to (+) or (-) to (-) bond, cannot be severed in the same electric dimension. This can only occur above 10 million K, which is a magnetic dimension

In stable mesons, the magnetic bond linking the two like charges is present and sufficient to hold the two components of the meson together at the A' midpoint. This magnetic bond creates a very stable meson structure, in contrast to the electric (-) 1/3 gluon to (+) 1/3 antigluon bond in the unstable mesons, which is easily severed leading to the destruction, or a major modification of the structural unit which it is a part; such as occurs in a proton or neutron decay process.

Synthesis

In the electric phase, six "color" up quarks, twelve "color" down quarks, six "color" antiup quarks, and twelve "color" antidown quarks were created, as previously described.

In addition, nine types of gluon/antigluon mesons were also produced and present in the E4 electric phase (Chart VIII, p.36-40).

In the following section we will concern ourselves only with

Precis : For pages 42 to 71

Outlined are the methods by which a gluon/antigluon meson^0, $(g/\bar{a})^0$, captures and bonds up quarks, $(U)^{(+)2/3}$; and down quarks, $(d)^{(-)1/3}$; to create two types of shell/g/a mesons. If the respective ratio is 2:1, a proton shell, $UUd^{(+)1}$, is formed and bonded to and surrounds a central g/ā meson, producing a proton shell/g/ā meson$^{(+)1}$, $UUd/g/\bar{a}^{(+)1}$; while if the ratio is 1:2, a neutron shell/g/a meson^0, $Udd/g/\bar{a}^0$ is created.

The proton shell, $UUd^{(+)1}$, and the neutron shell, Udd^0, are each internally bonded together using magnetic and electric charges; the sum total of both types of charges in each respective type of shell is called the weak force. The g/ā meson is also bonded together by it's magnetic and electric charges, the sum total of both types of charges therein is referenced as the strong force. The weak force also bonds the shell unit to the g/ā meson^0; while the g/ā meson's strong force bonds it to the shell unit.

If a single antiup or antidown quark, $\overline{U}$ or $\overline{d}$ inadvertently substitutes for a U or d in either a UUd or a Udd shell, a fractional charged unit will result. These fractional charged units are unstable in an electric or neutral-electric dimension; and thus will cause a shell containing one of them to immediately disassociate into their component quarks. Therefore, only proton or neutron shells which are whole numbers, (+1), (-1), or (0), can be created and remain stable, in an electric or neutral-electric dimension.

The exact same parallel holds for the antiproton shell ($\overline{UUd}$) and the antineutron shell ($\overline{Udd}$) forming around a g/a meson. If a single up or down quark, U or d, becomes a unit in either triad, it will also disassociate. Therefore, only antiproton shell/g/ā mesons and antineutron shell/g/ā mesons can be created. Both of these, however, will quickly disappear from the universe, as they will be converted into protons, electrons, positrons, neutrons and antineutrons by the action of positrons.

Each proton shell or neutron shell, UUd or Udd, enclosing a g/ā meson has a juncture point designated as A; while the enclosed g/ā meson also has it's own juncture point, designated as Á. Stable A junctures result from $(+)^{1/6}$ to $(+)^{1/6}$, or $(-)^{1/6}$ to $(-)^{1/6}$, magnetic bonds being formed between quarks therein. All $(+)^{1/6}$ to $(-)^{1/6}$ A bonds are unstable in magnetic dimensions and are designated as "0". The same rules and designations apply to the g/ā meson's point Á.

In the process which creates either a proton or neutron shell/g/ā meson, a bond which forms between like charges on both A and Á, a (+) to (+) or (-) to (-) bond, produces a completely stable unit; while a bond where A or Á is 0 results in a unit being unstable; which at some time thereafter will undergo either a proton or neutron decay process, thereby producing a respective neutron shell $(Udd)^0$, a neutrino and a positron; or a respective proton shell $(UUd)^{(+)1}$, an electron and an antineutrino, as the signature of the respective proton or neutron decay process.

Thus, there are extant **two types of protons**: a proton shell/g/ā meson, UUd/g/a $^{(+)1}$ and the proton shell/ $(UUd)^{(+)1}$; as well as **two types of neutrons**; a neutron shell/g/ā meson^0, Udd/g/ā^0 and a neutron shell $(Udd)^0$.

The g/a meson acts as if it were a wheel rim in each of it's respective shells, giving each of them great stability. Free standing shell units, however, either $(UUd)^{(+)1}$ or $(Udd)^0$, stripped of their g/a/ mesons, can momentarily have one of their bonds severed, due to their greater pliability without the stabilization produced by the g/a meson. This can cause radioactive decay, resulting in alpha particles, etc., when such stripped shell units are in certain positions in various elements.

These two respective proton and neutron types, along with the single type of electron, are the five bedrock entities from which our present Universe is constructed.

The unique systems in which these are synthesized are a magnetic phase of eleven dimensions in which the reaction rule is like attracts, opposites repel; while an electric phase of eleven dimensions which follows has a reverse rule: opposites attract and like repel.

It is these seemingly counterintuitive rules which permit the creation of a vast array of successive complex entities, composed of just two types of photons the P and the V. The existent universe therefore can then be seen as a sea of P and V photons, which have been clumped together in relative different packets, to produce a vast number of unique elements; then compounds of these elements; and finally into assemblages of these referenced as a sun, a planet, a moon, dust or us.

the six up quarks, twelve down quarks and nine types of mesons, which together generate many types of proton shell/(g/ā) mesons and neutron shell (g/ā) mesons.

Proton Shell/g/ā Mesons and Neutron Shell/g/ā Mesons

One of the two routes for the synthesis of the proton shell/(g/ā) meson or the neutron shell/g/ā meson is as follows: as soon as one of the nine types of (g/ā) mesons is created by the interaction of an electron and a positron, it acts as if it were an attractive nucleus. For example, if it first pulls and bonds an up quark ($U^{(+)2/3)}$) to its g side, which is at (-)1/3, it then must secondarily bond a down quark ($d^{(-)1/3)}$) to its ā side, which is at (+)1/3, both of which actions follow the opposites attract rule in effect in the electric phase; thereby forming a combined Ud complex with a net charge of (+) 1/3. The Ud complex will then seek out and merge with a third quark: either another up quark (U), thereby producing a proton shell (UUd) around the now central g/ā meson nucleus (UUd/g/ā $)^{(+)1}$; or it will add another down quark (d), producing a neutron shell (Udd) around its central meson nucleus (Udd/g/ā $)^{0}$. The up quark merger thus creates a (+) 1 charged proton shell; whereas the down quark merger creates a zero charge neutron shell, both centered around a now fully integrated g/ā meson nucleus, which is at zero (0) charge.

A (+) 1 or a zero (0) charged three-part shell structure, such as a respective proton or neutron shell, is extremely stable. A three-part shell structure, on the other hand, if formed with a fractional charge, is extremely unstable, and instantly disintegrates back into its separate quark components, leaving the g/ā meson free to begin its quark addition process once more.

Consequently, only combinations of up and down quarks can form a respective (+) 1 or zero charged three-part proton or neutron shell. The substitution of a single antiup or antidown quark in such a three-part structure would produce a fractional charged shell structure, which would quickly break down into its component quarks and g/ā meson.

The same is true in reverse: A respective (-) 1 or zero charged antiproton or antineutron shell can only be made up of antiup and antidown quarks. The substitution of a single up or down quark in such a three-part shell structure would produce a fractional charge unit, which would disintegrate into its component quarks and g/ā meson.

Now, let us turn to the creation of a proton shell /g/ā meson, whose entire process of generation is graphically portrayed in Charts IX, 1-6, p.43-46. Here, Number 1 (p.43), shows us a g/ā meson that has been formed by the interaction of an electron and a positron. It has a (g) gluon side at (-) 1/3 charge and an (ā) antigluon side at (+) 1/3 charge. Also shown is its overall net zero electric charge (0),

Chart IX

Proton Shell/g/ā Meson $^{(+)1}$ Synthesis

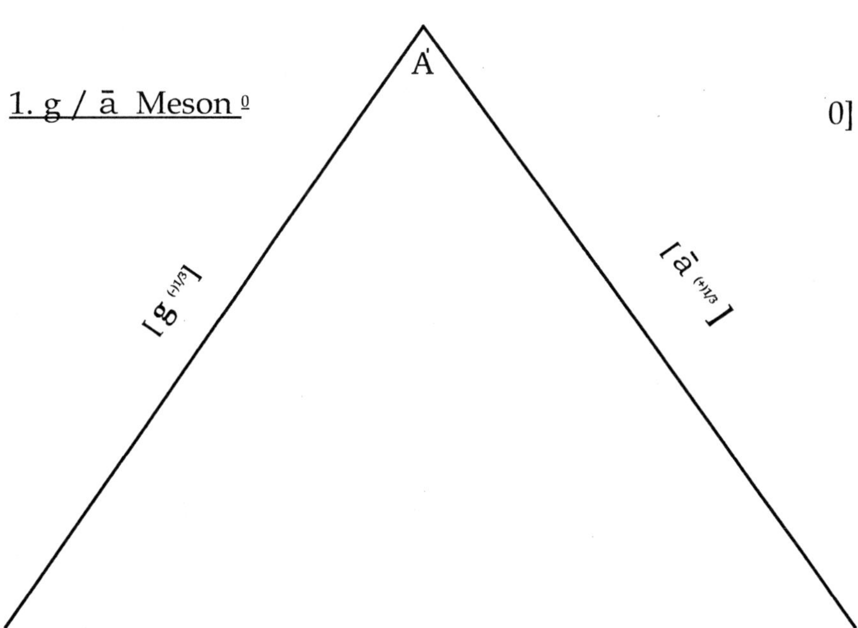

Chart IX (continued)

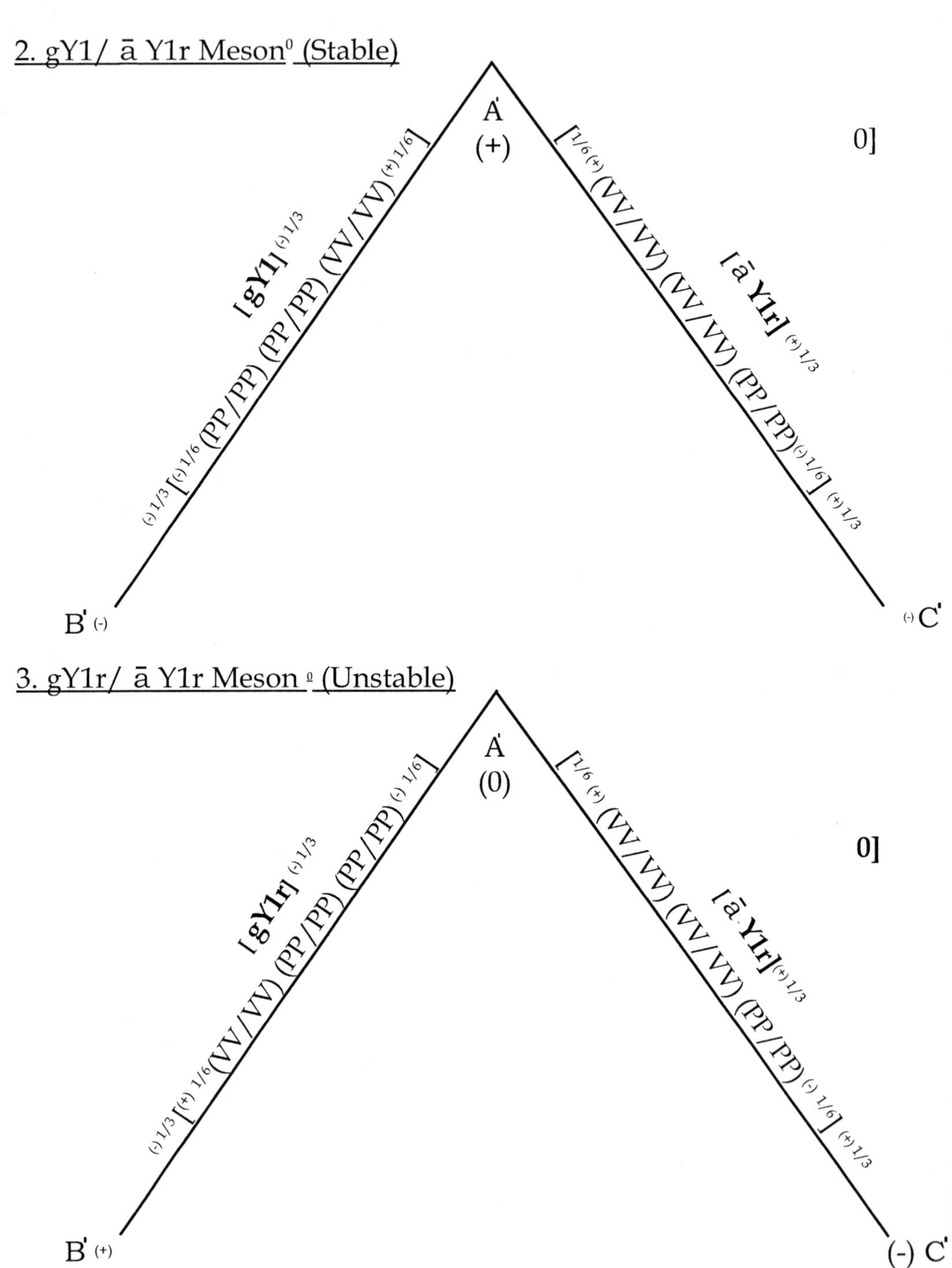

Chart IX (continued)

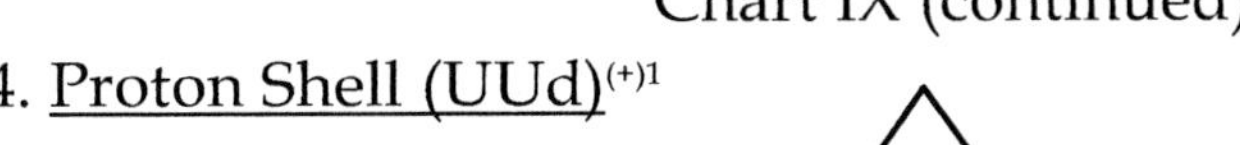

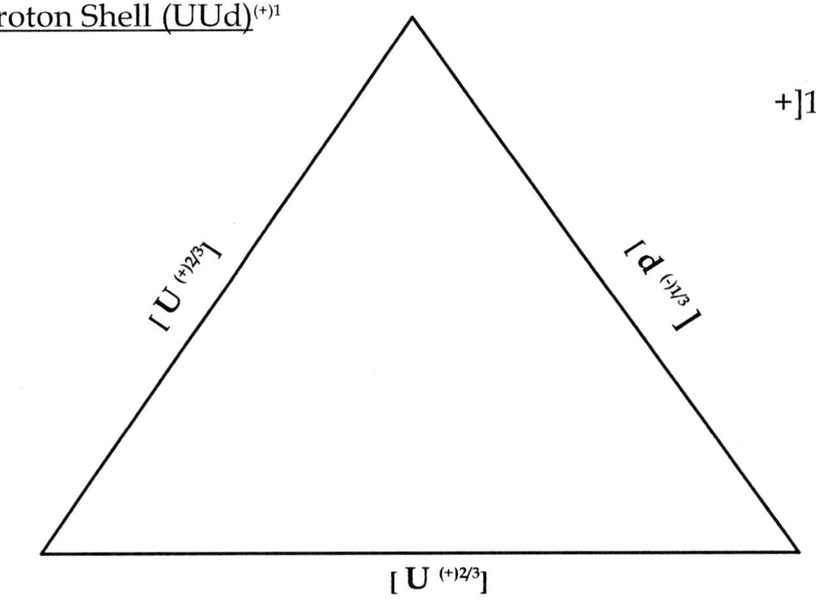

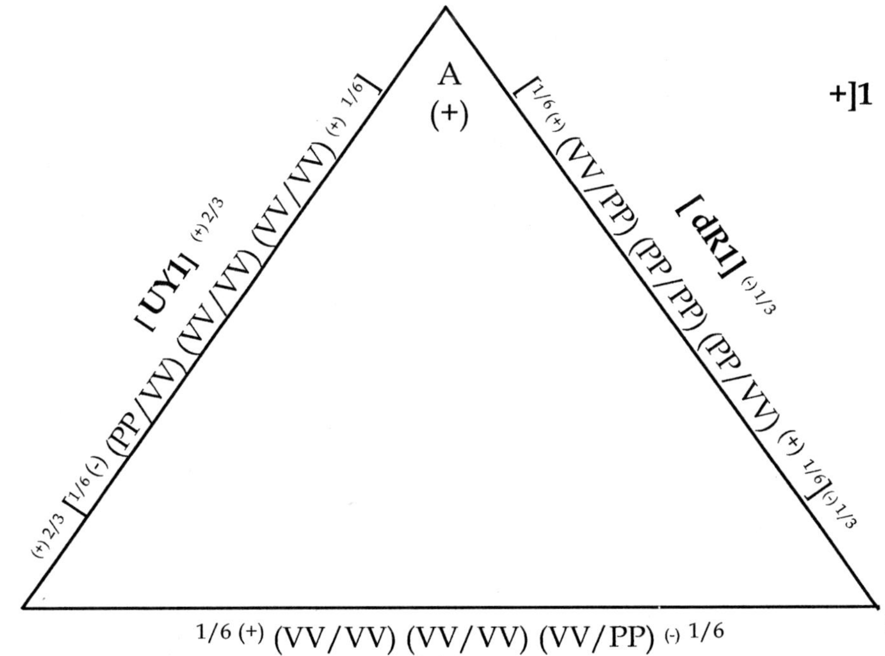

Chart IX (continued)

6. <u>Proton Shell [UR1/UR2/dB3]</u>$^{(+)1}$

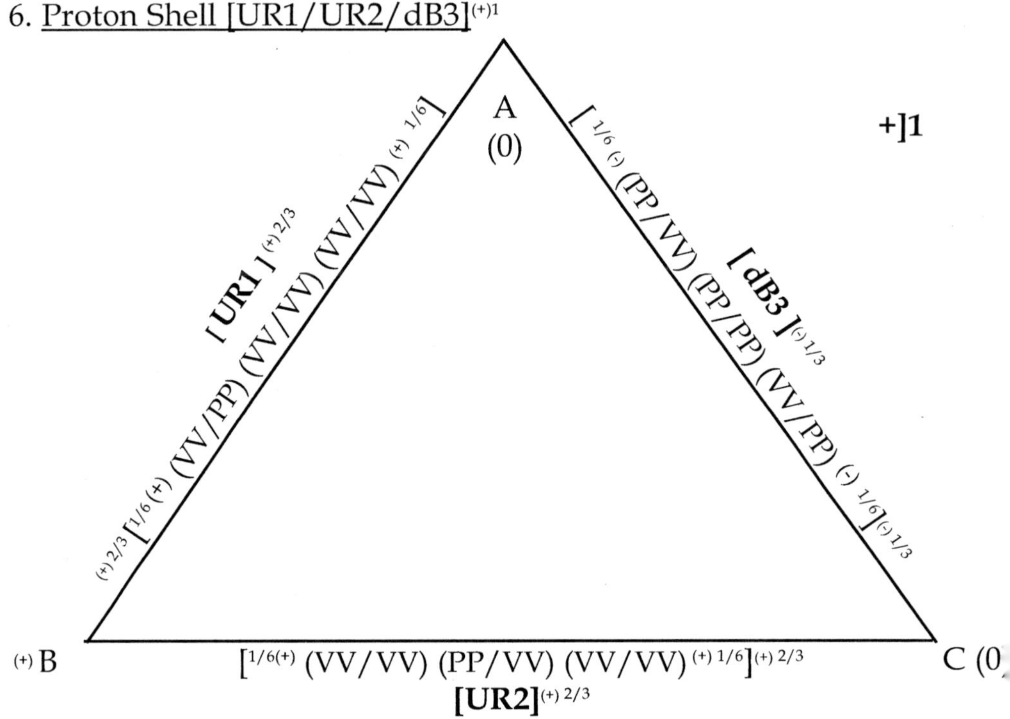

In Number 2, p.44, a yellow gluon, gY1, has been substituted for the g, and a yellow antigluon, ā Y1r, for the ā in Number 1 (also see Chart VIII, S1, p.36.) This produces at midpoint (Á) a (+) 1/6 to (+) 1/6 (+) bond, which is stable in magnetic and electric phases.

Number 3, in the Chart IX series, p.44, however, is an unstable meson, since it cannot unite at its midpoint (Á) a (+) 1/6 to (-) 1/6 charge into a magnetic bond in a magnetic domain, due to the opposites repel rule in effect in a magnetic phase.

The meson in Number 3, however, will be formed and held together by its (-) 1/3 to (+) 1/3 electric bond, generated by its respective g and ā gluons, upon cooling to below 10 million K. in the E4 electric dimension; but is vulnerable to dissolution by a reactive particle or energy flux.

Number 4, p.45, above, shows a proton shell, UUd$^{(+)1}$, which can form around a g/ā meson nucleus to produce a proton shell/g/ā meson. In addition, all of the U and d units are identified with respect to their intrinsic charges, (+) 2/3 and (-) 1/3, respectively; as well as the overall net charge displayed by the unit, (+) 1.

Number 5, p.45, substitutes UY1 and UY1r for the two up quarks and dR1 for the down quark, in the previously shown Number 4; thereby creating a proton shell with a (+)(+) stable magnetic bond at its midpoint A, indicated by a (+).

The critical position in Number 5 is A, with its operative (+) magnetic bond, which is then able to magnetically bond in a magnetic phase interaction, with a same type (+) center Á bond of a (+)(+) type of a g/ā meson it has formed around. Therefore, the g/ā meson in Chart IX, Number 2, p.44, would produce a stable proton shell/g/ā meson with the proton shell Number 5, Chart IX., p.45. It could resist nearly all but the most vigorous reactive particles and their interactions.

Chart IX, Number 7, p.48, depicts the proton shell/g/a meson which incorporates the proton shell in Number 5, around the g/a meson exhibited in Number 2, p.44. The (+) at the A of the proton shell can form a stable bond with the (+) at the Á of the g ā meson in a magnetic dimension.

Chart IX, Number 6, p.46, on the other hand, displays an unstable proton shell with a (+) to (-) midpoint A, designated by a [0] underneath. When it forms around a stable or unstable meson, it cannot form an A to Á bond and is, therefore always an unstable unit. Number 8, p.48, shows this type of proton shell with its (+) to (-) midpoint A unable to bond with the (+)(-) midpoint Á meson. This type of proton shell/g/ā meson is, however, held together by its (+) 1 overall electric charge in an electric phase. But the stability in terms of holding the meson securely within the boundary of the proton shell is very tenuous, producing a relatively unstable unit, subject to a proton decay sequence under appropriate circumstances.

In fact, a slight energy flux can cause the meson to be dislodged out of the proton shell in Number 8, thus initiating a proton decay process, which we will describe later. This cannot

Chart IX (continued)

7. Proton Shell [UY1/UY1r/dR1]/g/ ā Meson [gY1/ ā Y1r] (+)1 (Stable)

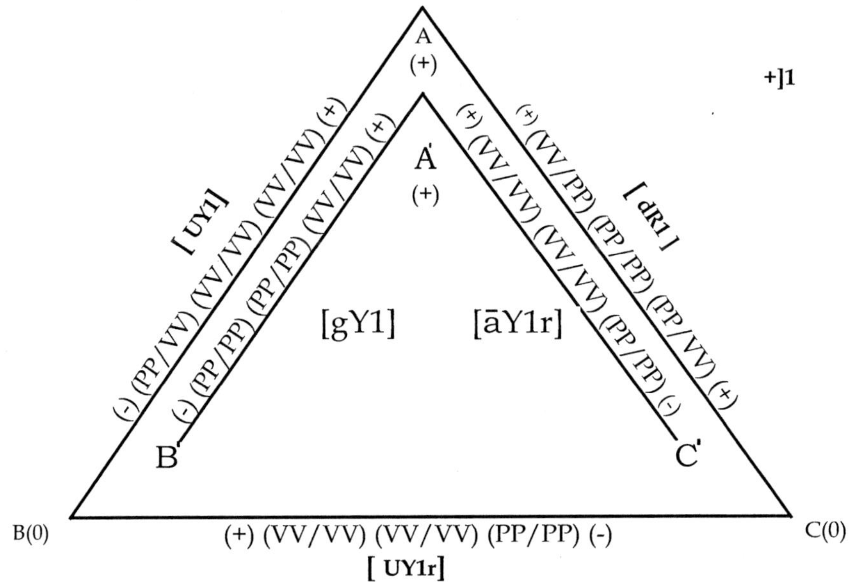

8. Proton Shell [UR1/UR2/dB3]/g/ā Meson [gY1r/ ā Y1r] (+)1 (Unstab

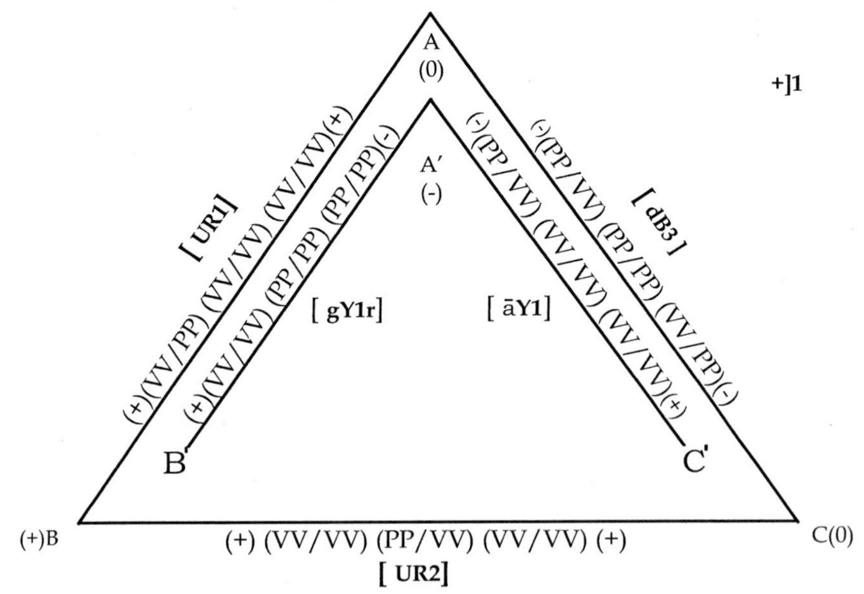

happen to Number 7, p.48. But, first, let us discuss some insights into the nature of magnetic bonds and electric bonds.

Magnetic Bonds and Electric Bonds

The (+) 1/6 magnetic charge on a (+) monopole, $(VV)^{(+)1/6}$, or (-) 1/6 magnetic charge on a (-) monopole, $(PP)^{(-)1/6}$, is created when two (+) 1/12 V photons or two (-) 1/12 P photons, respectively, bond with each other in a magnetic phase, where the rule is like attracts, opposites repel. These two types of monopoles are the basic magnetic structures that permit the formation of all higher electric charged structures.

When two like charge monopoles are brought close together, they establish a new type of bond and a new type of additive charge between them. For example, two (+) 1/6 charges add their charges into a (+) 1/3 charge level that is unlike the two separate (+) 1/6 charges in properties.

Due to this, the (+) 1/3 charge level is designated as electric. The same also holds true for the (-) 1/3 charge level, the result of combining two (-) 1/6 charges.

When a (+) 1/3 and a (-) 1/3 charge combine, following the opposites attract rule in an electric phase, they also add their charges, which results in a zero (0), neutral, electric charge.

Other electric charges are the (+) and (-) 2/3, (+) and (-) 1.

These electric charges are all, however, based on having multiple (+) or (-) 1/6 monopoles present. For example, in a (+) 1 positron, six (+) 1/6 charge monopoles are present in three pairs of two each. Each pair is called a W+ boson, which displays a (+) 1/3 charge, thus requiring three such W (+) 1/3 bosons to be present in a positron in order to add up to (+) 1.

Yet, under certain conditions in three related elements; iron, cobalt and nickel, referred to as the ferromagnetic elements, specifically when they demonstrate magnetic properties, it is understood to be as a result of all of their electric bonds in their protons and neutrons being severed and turned off, leaving only (+) 1/6 and (-) 1/6 magnetic charges present and operative; thereby producing magnets.

The "magnetic tug" between (+) and (-) poles produced by these elements is the result of respective (+) or (-) 1/6 charges at each respective pole.

The (+) and (-) 1/12 photons that compose each respective (+) and (-) 1/6 monopole were created from the substance of their respective V_E or P_E gravitons, when each of these two gravitons converted some of their gravitational energy substance, E, into mass-matter substance, M, via the $E=MC^2$ process, explained in much greater detail in later chapters. The V_E or P_E graviton substance is not, however, composed of V or P photons, respectively. Each of their gravitational substances is unique.

When the north pole of a bar-magnet, projecting a homogenous grouping of (+) 1/6 magnetic charges, comes in close proximity to the south pole of another bar-magnet displaying a homogenous grouping of (-) 1/6 charges, a "tug" is felt up to the moment of contact. At contact, a zero magnetic, or neutral charge is effectuated which holds both poles together. When they are "tugged" apart, both poles regain their respective (+) 1/6 and(-) 1/6 magnetic charges, respectively the north and south poles of the bar-magnet.

These (+) and (-) 1/6 magnetic charges are, however, operating in a partial electric and partial neutral electric domain, due to the earth being both an E11 (electric) and an N11 (neutral-electric) dimensional domain, simultaneously. As magnetic charges move through these electric spaces, they must obey the charge rules in effect in an electric domain; which requires like charges repel and opposites attract. Therefore, the (+) and (-) 1/6 magnetic charges attract, instead of repelling each other as would be the rule in their originating magnetic domain, the M4 dimension; thus, the "tug" between the opposite magnetic poles (north and south) on earth.

Proton Shell /g/ā Meson Decay Process

The proton shell /g/ā meson, $(UUd/g/\bar{a})^{(+1)}$, decay process transforms it into a neutron shell $(Udd)^0$, a neutrino (V_E) and a positron (e+).

Of the numerous types of proton shell /g/ā mesons that are created, those in which the A position in the proton shell and the Á in the g/ā meson cannot form a magnetic bond between themselves are the types that mainly become subjects of the decay process (Chart IX.8), p.48. If, on the other hand, a bond can be formed between A and Á, the structure is stable and usually not subject to the decay process. The former we classify as unstable, and the latter as stable proton shell /g/ā mesons. An example of the latter is shown in Chart IX.7, p.48. Here, Á at (+) and A at (+) can and do, in a magnetic domain, bond together due to the like-attracts-like rule therein.

Chart X.1

Proton Shell ((UR1)(UR2)(dB3)/g/ā Meson (gY1r)(ā Y1))$^{(+)1}$(Unstable)

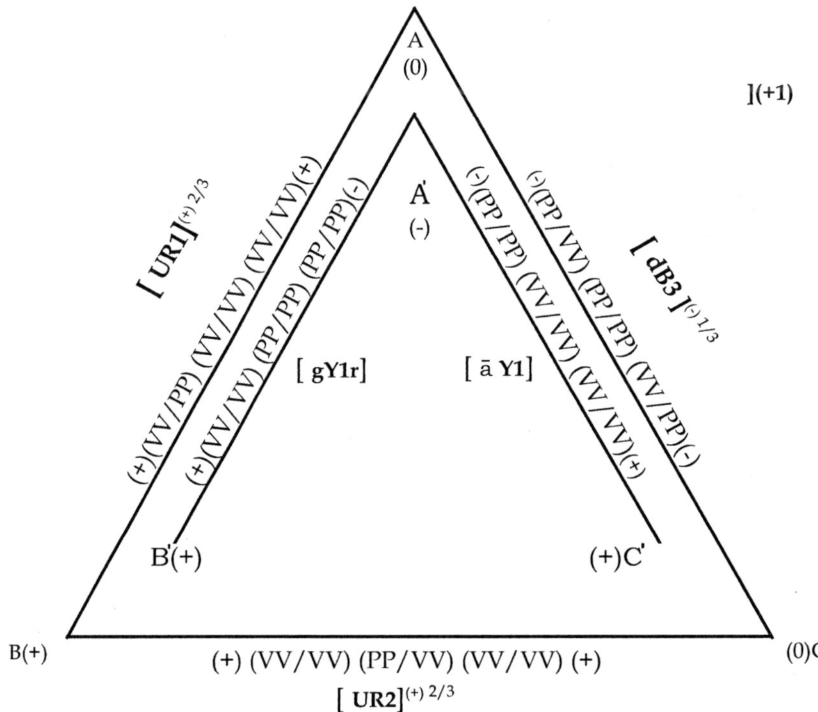

Chart X.2

A(0)
PROTON
SHELL
(ΔABC)
[UR1]
(+)(VV/PP) (VV/VV) (VV/VV)(+)
(-)(PP/VV) (PP/PP) (VV/PP)(-)
[dB3]
[UR2] (Shared Side)
B(+)
(+) (VV/VV) (PP/VV) (VV/VV) (+)
(0)C
[gY1r]
(+)(VV/VV) (PP/PP) (PP/PP)(-)
g/ā
MESON
(ΔA'BC)
[āY1]
(-)(PP/PP) (VV/VV) (VV/VV)(+)
A'(-)

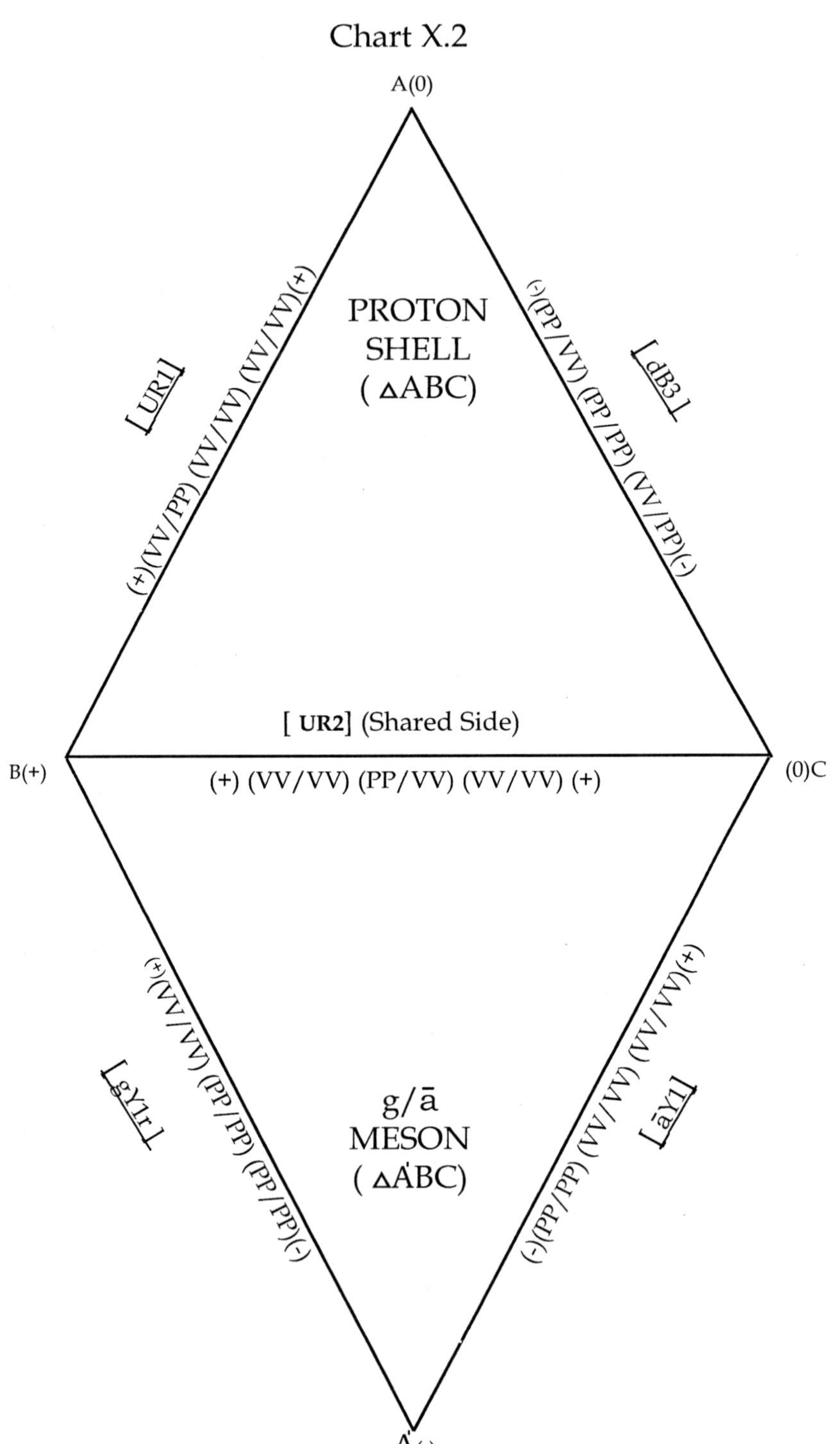

Unstable Proton Decay Process

An example of an unstable proton shell /g/ā meson is illustrated in Chart X.1, p.51. This is due to the Á position in the g/ā meson being a (-) and the A position in the proton shell being a (0); the "0", a result of having (+) and (-) charges at its juncture, cannot form a bond in a magnetic dimension. Therefore, it is a prime example of a proton shell /g/ā meson that can undergo a decay process.

An energy flux can cause the g/ā meson in Chart X.1 (p.51) to fall out of its proton shell and form a double triangle structure, as seen in Chart X.2 (p.52).

Here, the proton shell (△ABC) now shares its BC bottom side (UR2) with the out-fallen two sides of the g/ā meson. The meson thereby forms an △ÁBC triangle structure, connecting the g/ā meson's two side arms, ÁB (gY1r) and ÁC (ā Y1), and the BC bottom side (UR2) of the proton shell (△ABC).

The g/ā meson, as two sides of the △ÁBC triangle, is highly unstable, due to it having to share its BC side with the proton shell (△ABC). A nearby positron, for example, can attract the (-) 1/3 gY1r gluon and tear it out of its bond with the (+) 1/3 ā R1 antigluon due to its greater (+) charge of (+) 1.

As a result, the g/ā meson triangle (△ÁBC) is immediately torn apart and converted into three new structures: an electron (e-), a positron (e+) and an up quark (U), as seen in Chart X.3,p.54.

It does this, however, in an unusual way. The electron (e-) is produced by assembling three W- bosons (PP/PP), taking two from the ÁB side (gY1r) and one from the ÁC side (āY1). The resultant electron triangle is labeled e-.

The positron, labeled as the e+ triangle, is created from the two W+ bosons (VV/VV) on the ÁC side of the āY1 and the contiguous single W+ boson (VV/VV) on the BC right side, at the bottom of the proton shell, (the UR2, shared side.)

What remains is an up quark on the left side of a △ÁBC triangle. It is labeled U, composed of the W+ boson (VV/VV) remaining on the ÁB side of the former g/ā meson and the two boson remnant of the BC side of the proton shell, (VV/VV)(PP/VV), UR2, shared side.

Chart X.3

A(0)
PROTON
SHELL
(ABC)
[UR1]
(+)(VV/PP) (VV/VV) (VV/VV)(+)
(-)(PP/VV) (PP/PP) (VV/PP)(-)
[dB3]
[UR2] (Shared Side)
B
C
(VV/VV)
(PP/VV)
(VV/VV)
U
(up quark)
e+
(positron)
(VV/VV)
[gY1r]
(PP/PP) (PP/PP)
e-
(electron)
(VV/VV) (VV/VV)
[āY1]
(PP/PP)
A′

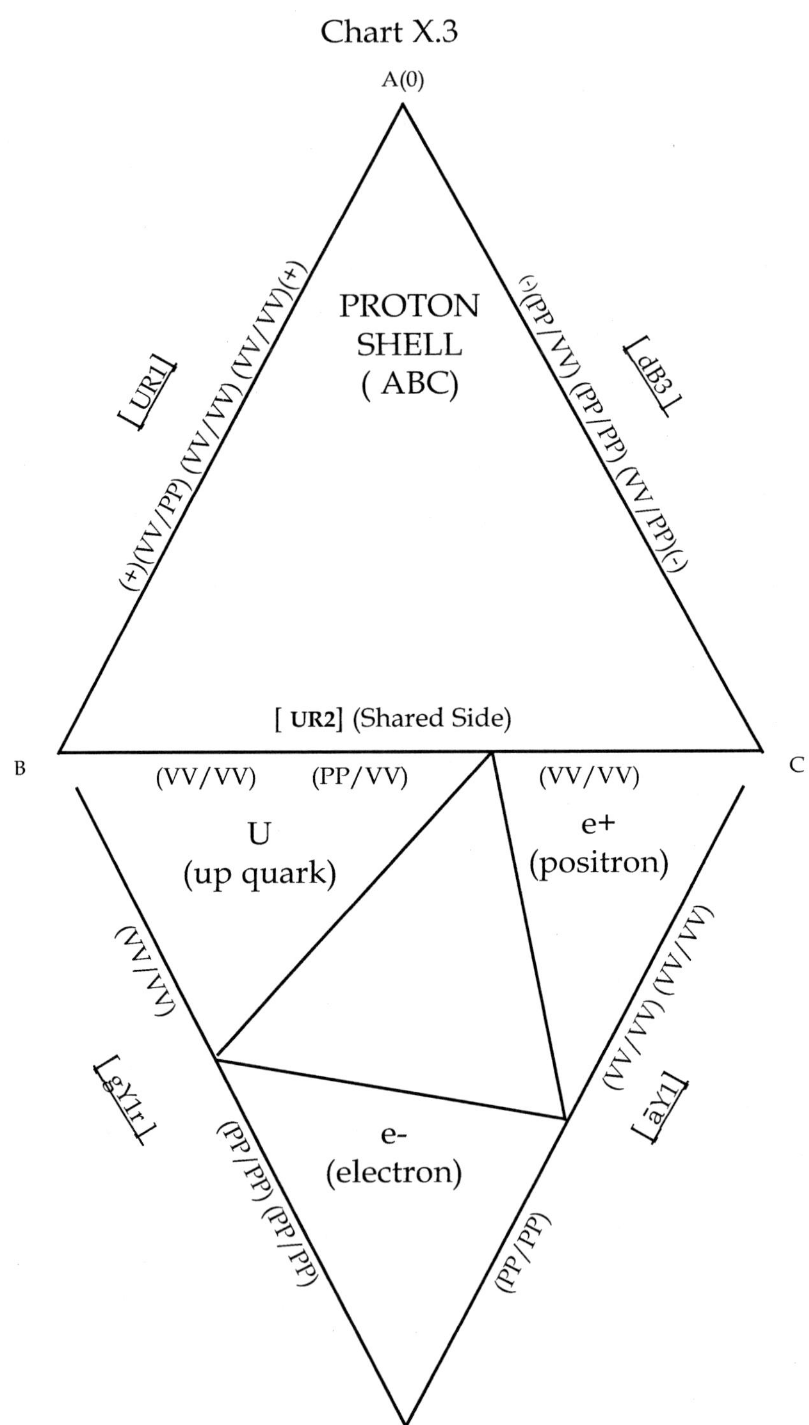

Chart X.4

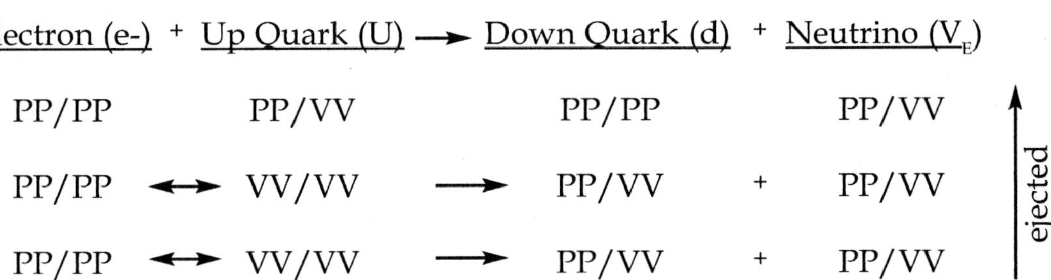

Chart X.5

Neutron Shell (UR1/dB3/dY1)

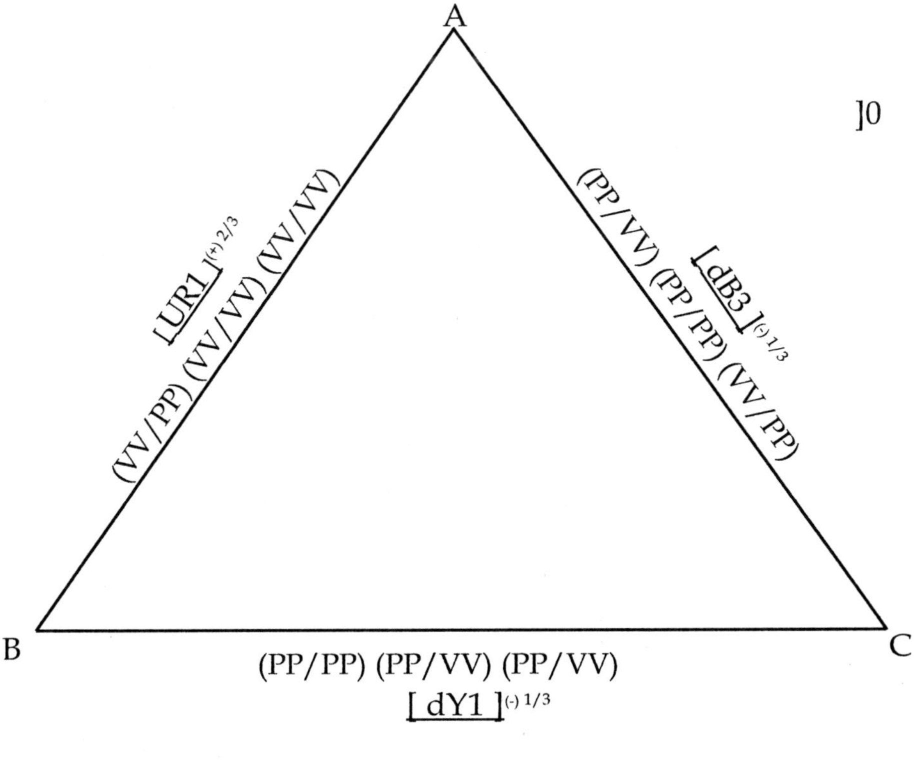

A

]0

[UR1]$^{(+)2/3}$

(VV/PP) (VV/VV) (VV/VV)

(PP/VV) (PP/PP) (VV/PP)

[dB3]$^{(-)1/3}$

B

C

(PP/PP) (PP/VV) (PP/VV)

[dY1]$^{(-)1/3}$

The next step in the decay process is the ejection of the positron (e+) outside the area of the proton shell /g/ā meson. This, then, causes the up quark (U) and the electron (e-), which are contiguous to each other, to interact, as described in Chart X.4, p.55.

The reaction in Chart X.4,p.55. is an unusually violent one. It causes the two W-bosons (PP/PP) in the electron (e-) and two W+ bosons (VV/VV) in the up quark (U), both shown interacting (◄►), to be torn apart into their monopole constituents, four (PP) and four (VV), which are then bonded into four Z bosons (PP/VV). Two of these Z bosons (PP/VV) then attach themselves to the remaining W- boson (PP/PP), left over from the former electron (e-) in Chart X.4, p.55; thereby creating a down quark, d, ((PP/PP)(PP/VV)(PP/VV)), shown after the reaction arrow (⟶); while the remaining two Z bosons (PP/VV), also formed in Chart X.4, attach themselves to the Z boson (PP/VV) remaining from the up quark (U) which initially reacted with the electron (e-) in Chart X.4. This produces a neutrino, V_E , (PP/VV)(PP/VV)(PP/VV), which is ejected into the adjacent space. The down quark, d, a dY1 type, is then immediately snapped between the two remaining sides of the former proton shell, AB (UR1) and AC (dB3), becoming the third side, BC (dY1), to produce a neutron shell, AB(UR1)AC(dB3)BC(dY1), depicted in Chart X.5, p.55.

In summary, the proton shell /g/ā meson (in Chart X.1, p.51) decayed into a neutron shell, a positron and a neutrino; the latter two being ejected into the surrounding space (Chart X.5).

In this rather complicated process, we began with thirty-six V photons and twenty-four P photons composing the proton shell /g/ā meson structure in Chart X.1. At the conclusion of all of the reactions, we still have the same respective number of V and P photons, 36 and 24; but now dispersed into three different structural entities: a positron (12 V photons), a neutrino (6 V photons and 6 P photons) and a neutron shell (18 V photons and 18 P photons). Thus, both sides of this reaction equation are balanced.

Neutron Synthesis

In a previous section dealing with proton synthesis, we described the electric phase system that produced various types of g/ā mesons as a result of a reaction between an electron and a positron. These mesons, in turn, attracted and bonded an up quark, and then a down quark, in a manner as if the g/ā meson were an attractive nucleus.

These two types of quarks (U and d) and the g/ā meson together produce a combined (+) 1/3 complex that will, in turn, react with a third quark member.

If that member were another up quark, a proton shell, $UUd^{(+)1}$, would thus form around the g/ā meson nucleus, producing a proton shell/g/ā $meson^{(+)1}$. If, on the other hand, a down quark were the third member, then a neutron shell Udd^{0}, would be in place around the g/ā meson, creating a neutron shell/g/ā $meson^{0}$.

We will now explore this latter case.

Neutron Shell /g/ā Meson

We have already sufficiently explored the nine types of g/ā mesons that are produced via the electron and positron reaction. These same g/ā mesons also serve as the inner nucleus for a neutron shell, Udd^{0}, creating a neutron shell /g/ā meson, Udd/g/ā ^{0}.

Chart XI.1 (p.59) illustrates the electric components of a zero charged neutron shell.

Chart XI.2 (p.59) displays the neutron shell with a (-) charge at its A position, which has been determined by the specific two quarks (UY1r and dY3r), intersecting at A. The neutron and its g/ā meson will be stable if the Á position on the meson will also be a (-) charge.

Chart XI.3 (p.60) presents a different neutron shell which displays a zero (0) charge at the A position. The presence of this charge precludes any bond being formed with and between any charge on any g/ā meson's Á position. All neutron shell/g/ā mesons of this type are fated to be unstable and eventually to undergo a neutron decay process.

Chart XI.4 (p.61) displays a g/ā meson, (gY1r/ āY1), incorporated into the center of the neutron shell, (UY1r)(dY3r)(dY1r), which is exhibited in Chart XI.2, (p.59).

This is a stable configuration owing to the attractive magnetic bond between the respective A′ (-) and the A (-) positions on the g/ā meson and neutron shell.

Chart XI.5 (p.62) illustrates the structure of an unstable, or decay prone neutron shell /g/ā meson. This unstable condition is based on both the A and A′ positions on the respective neutron shell and the g/ā meson being both zero (0) charges, thus precluding any magnetic bond formation between them in a magnetic dimension, but allowing a (+) to (-) bond to be formed in an electric dimension.

A close inspection of Chart XI.5's unstable structure and the stable configuration depicted in Chart XI.4, (p.61), shows only one difference between them: the down quark (dY3r) in Chart XI.4, has been replaced by its 180° reverse structure, the down quark (dY3), shown below.

dY3r = (PP/VV)(PP/PP)(PP/VV)
dY3 = (VV/PP)(PP/PP)(VV/PP)

This illustrates how the very smallest of the structural difference can cause a stable construction to become unstable.

Neutron Decay

Neutron decay is a process in which a neutron shell/g/ā meson^0 is transformed into a proton shell (UUd), an electron (e-) and an antineutrino ($\overline{V}_E$).

The neutron shell /g/ā meson^0, depicted in Chart XII.1, (p.64) is a highly unstable structure due to the inability of its A (0) to A′ (+) magnetic bond to form, as well as the magnetic non-bonding of its B (0) to B′ (-) and C′ (0) to C′ (-) bonds. It is a prime candidate for decay.

In parallel to the proton shell /g/ā meson decay process we have previously illustrated, the g/a meson in Chart XII.1 also falls out of the neutron shell, due to the inability to form a magnetic bond, as shown in Chart XII.2, (p.65), and thus it forms a second

Chart XI.1

Neutron Shell (Udd)$^{(0)}$

[U$^{(+)2/3}$]
[d$^{(-)1/3}$]
[d$^{(-)1/3}$]
]0

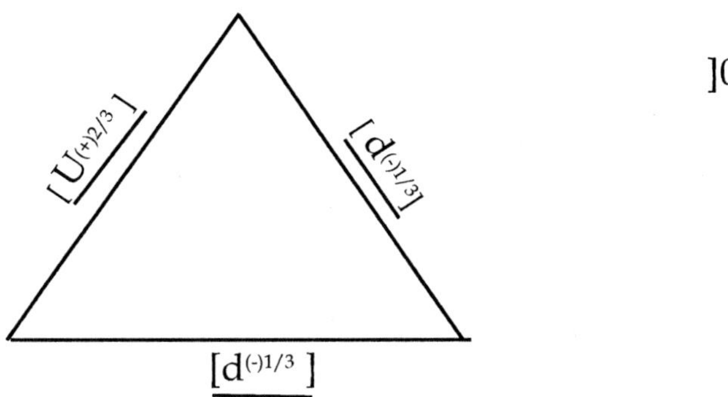

Chart XI.2

Neutron Shell ((UY1r)(dY3r)(dY1r))^{0}

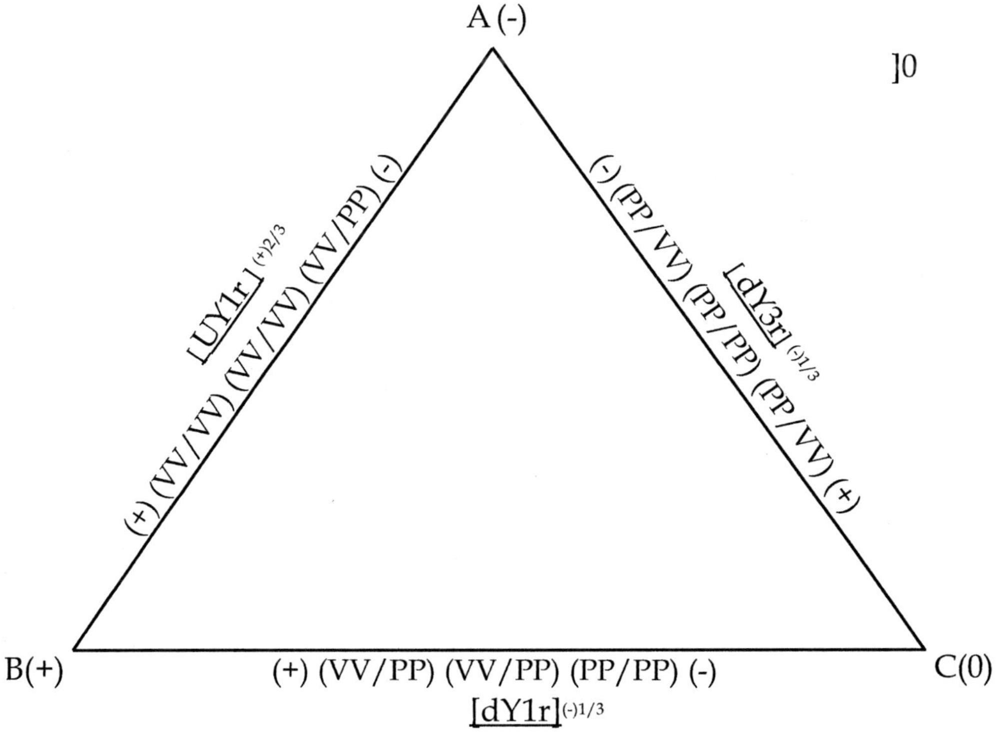

A (-)
]0
[UY1r]$^{(+)2/3}$
(+) (VV/VV) (VV/VV) (VV/PP) (-)
(-) (PP/VV) (PP/PP) (PP/VV) (+)
[dY3r]$^{(-)1/3}$
B(+)
C(0)
(+) (VV/PP) (VV/PP) (PP/PP) (-)
[dY1r]$^{(-)1/3}$

Chart XI.3

Neutron Shell ((UY1r)(dY3)(dY1r))

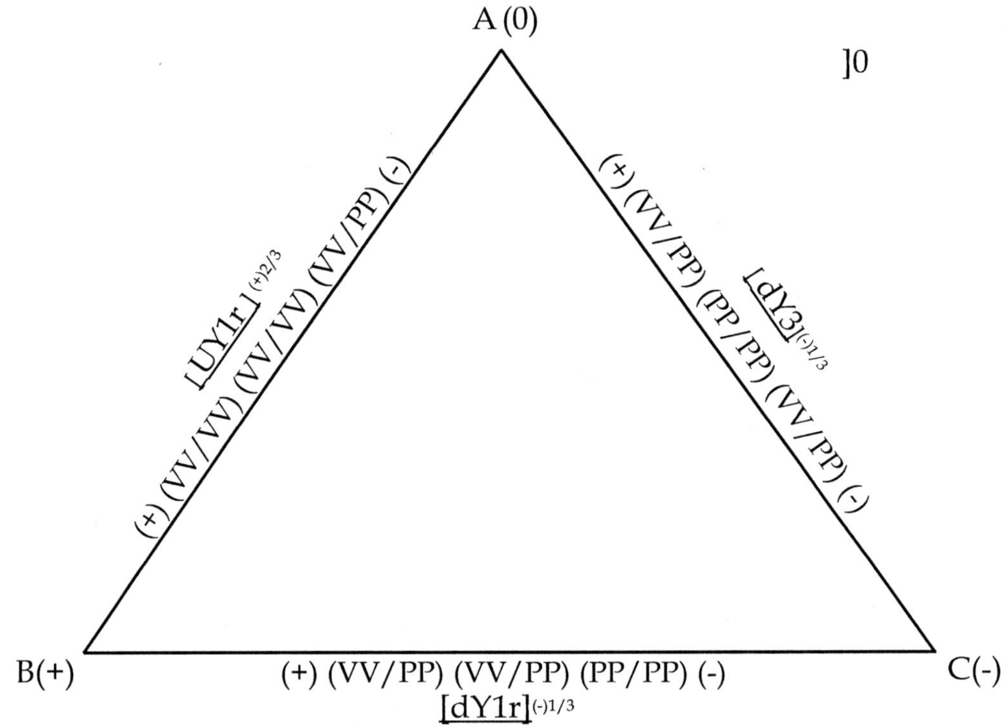

Chart XI.4

Neutron Shell ((UY1r)(dY3r)(dY1r)/g/ā Meson ((gY1r)(āY1)))⁰

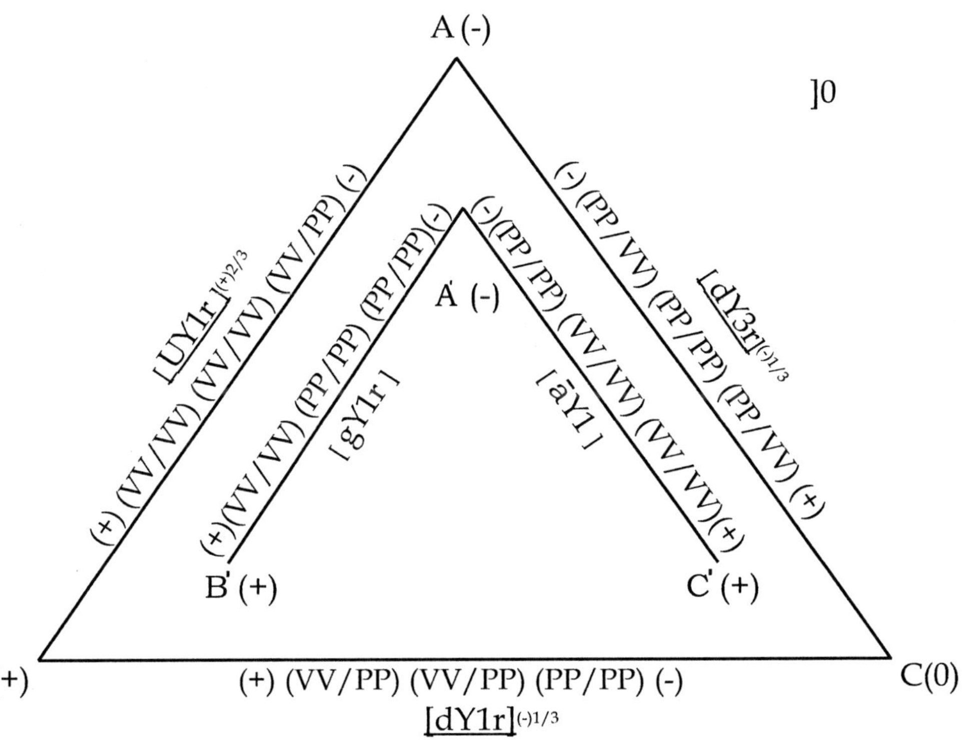

Chart XI.5

Neutron Shell ((UY1r)(dY3)(dY1r)/g/ā Meson ((gY1r)(ā Y1)))^0

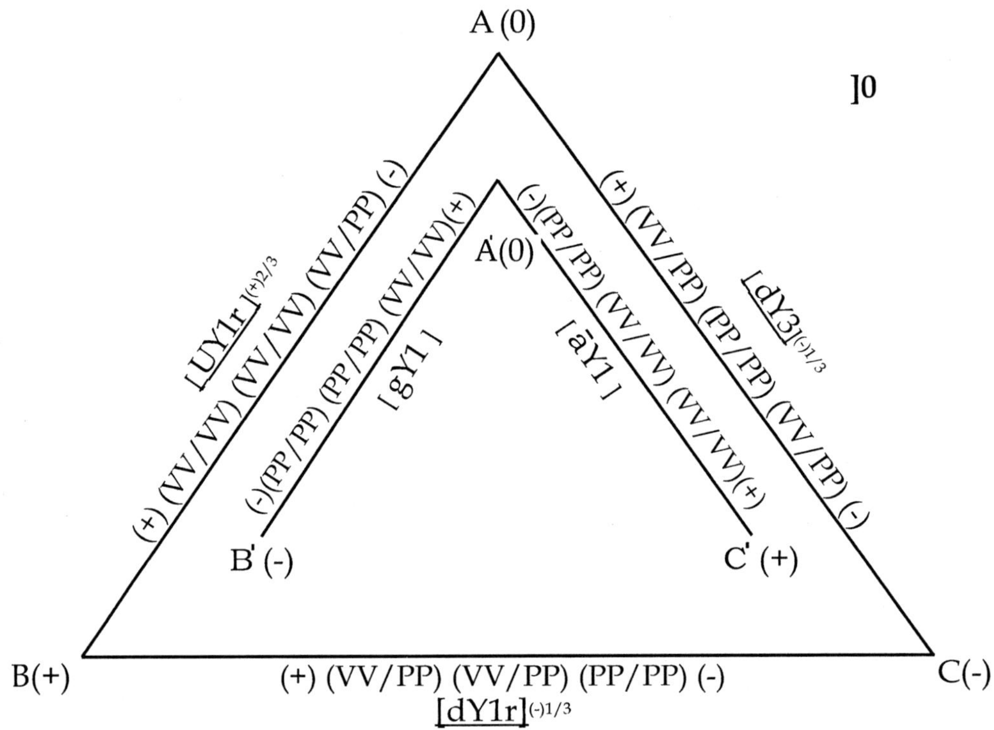

triangular structure (ÁBC) opposite the neutron shell (ABC). Both triangular structures, referenced in the last paragraph on p.58, now share the BC (dY1) side of the neutron shell.

Unlike the proton decay process where the g/ā meson fell out of the proton shell and was then active in a (+1) electric charge environment, in the neutron decay process at the same point (Chart XII.3), (p.66), the g/ā meson falls out into a zero electric zone (0), (△ÁBC).

A zero or (-) 1 electric charge zone favors activating only the P photon derived structures, whereas a (+1) environ allows only the V photon type structures to become active. In the proton decay process, the positron is ejected because it is too active to be held within the process dynamics, of a (+) 1 zone.

Here, in the neutron decay process, the newly formed, activated electron (e-) is ejected into the surrounding area due to its great reactivity within its favored neutral (0) environment.

The positron, now subdued by this same neutral (0) electric region, is not ejected, reacting instead with the d quark. This interaction, however, is not as violent as was the electron (e-) and the up quark (U) in the proton decay process, cited previously, where the monopole-to-monopole bonds in the electron and up quark were shattered producing free monopoles which then reformed as Z bosons (PP/VV).

The result of the positron (e+) and d quark reaction, however, is the shattering of only the boson-to-boson bonds holding both of these two units together, thus causing the formation of three free and unbonded bosons in each of these two reactants (see Chart XII.3A, p.67). All monopole to monopole bonds here are intact

Two of the three newly freed W+ bosons (VV/VV) from the positron (e+), will then merge with one of the two newly freed Z bosons (PP/VV) from the d quark; resulting in all three bosons then combining into the up quark, as shown in Chart XII.3A (p.67), $[(PP/VV)(VV/V)(VV/VV)]^{(+)2/3}$, a UR1r type, (third column).

Simultaneous within the aforementioned 3A process, the remaining free W+ boson (VV/VV) left over from the shattered positron (e+) then attaches itself to the remainder of the d quark, the W- boson (PP/PP) and the Z boson (PP/VV). These three bosons then combine into an antineutrino, $((PP/PP)(PP/VV)(VV/VV))^{0}$,

Chart XII.1

Neutron Shell ((UR1r)(dY1)(dB1))/g/ā Meson ((gY1)(āY1r))[0]

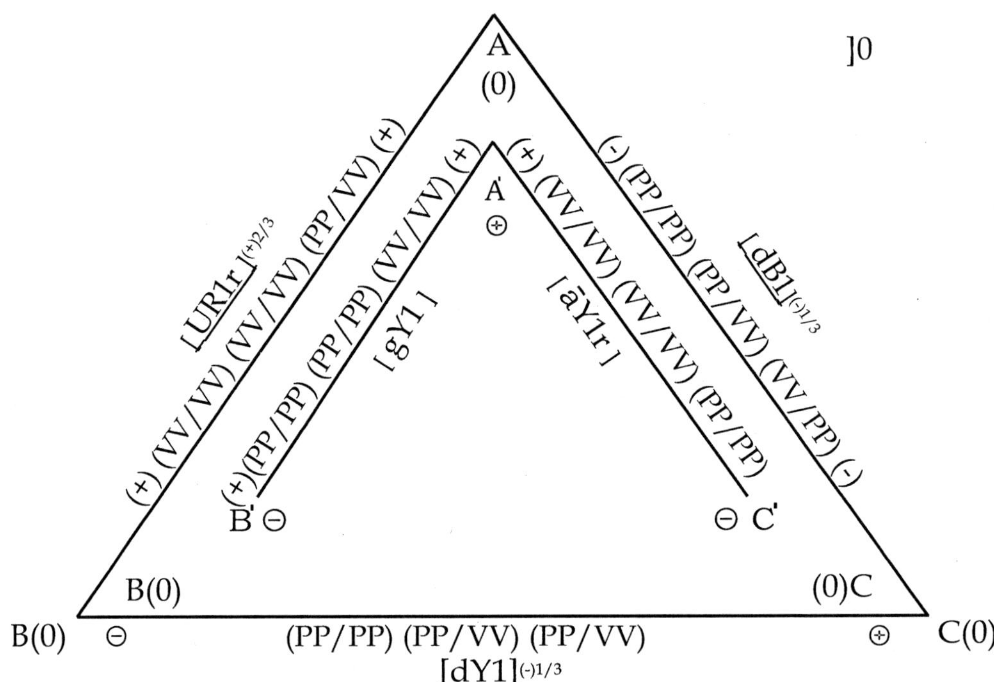

Chart XII.2

A
]0
(+) (VV/VV) (VV/VV) (PP/VV) (+)
[UR1r]
(-) (PP/PP) (PP/VV) (VV/PP) (-)
[dB1]
NEUTRON
SHELL
(ΔABC)
[dY1]
B
C
(-) (PP/PP) (PP/VV) (PP/VV) (+)
(-) (PP/PP) (PP/PP) (VV/VV) (+)
[gY1]
g/ā MESON
(ΔA'BC)
(+) (VV/VV) (VV/VV) (PP/PP) (-)
[āY1r]
A'

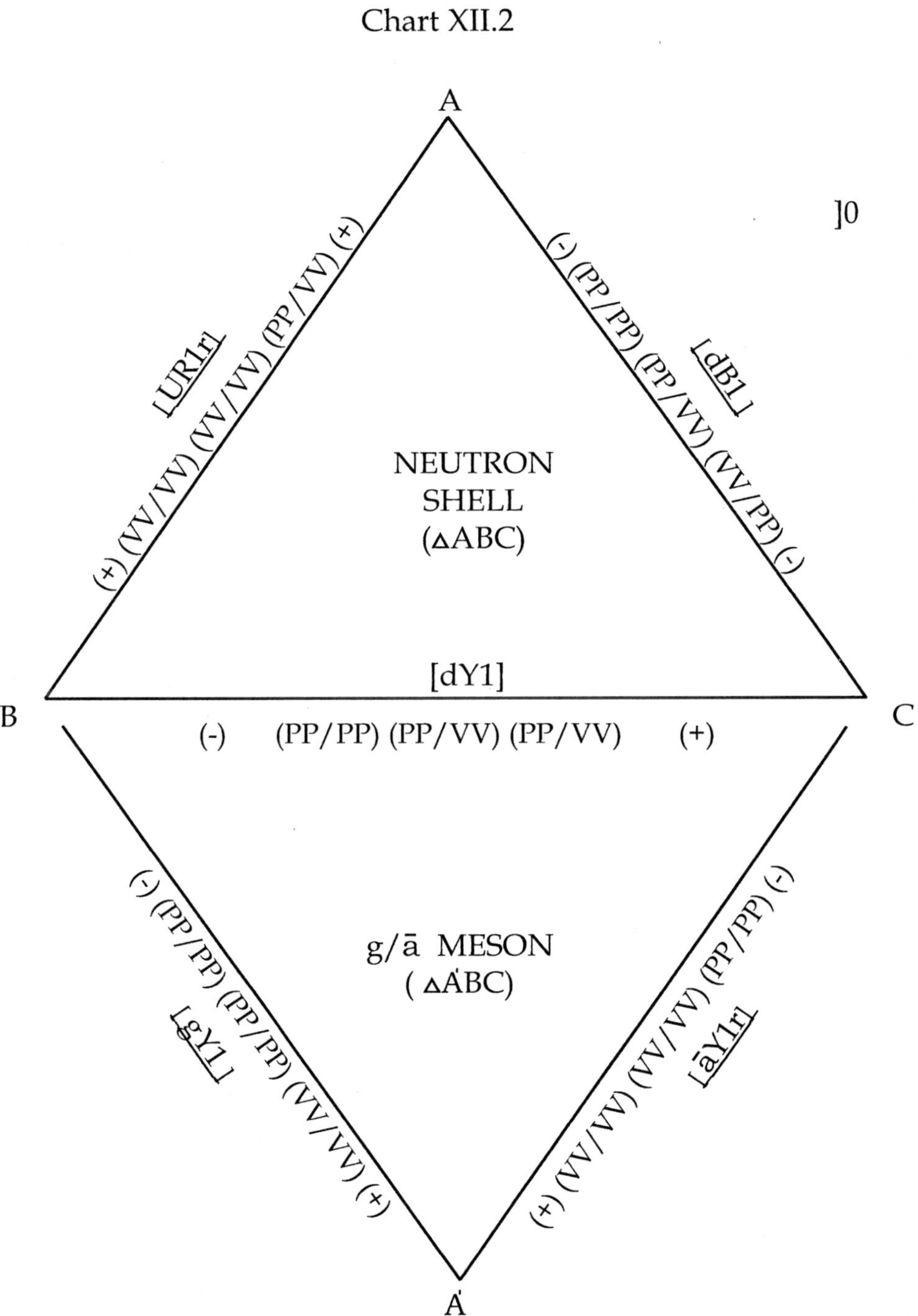

Chart XII.3

A(0)
]0
(+) (VV/VV) (VV/VV) (PP/VV) (+)
[UR1r]
(-) (PP/PP) (PP/VV) (VV/PP) (-)
[dB1]
NEUTRON
SHELL
[dY1]
B(0)
C(-)
(PP/PP)
(PP/VV) (PP/VV)
d
(d quark)
e-
(electron)
(PP/PP) (PP/PP)
[gY1]
(PP/PP)
(+) (VV/VV) (VV/VV)
[āY1r]
e+
(positron)
(VV/VV)
A′
(+)

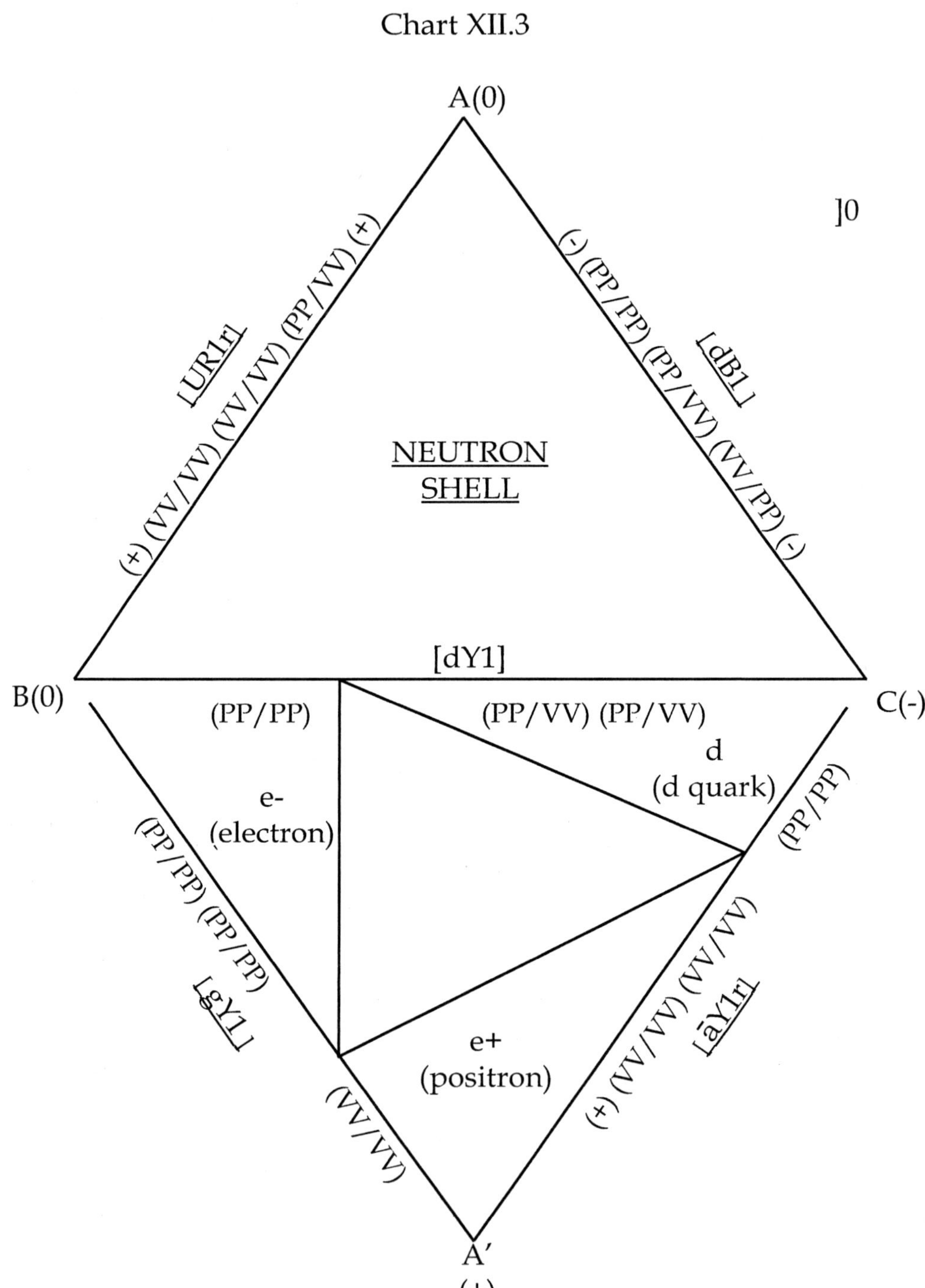

which is then ejected from the reaction area (↑).

As the up quark (UR1r), whose construction is depicted in Chart XII.3A (p.67) below, was completed, it was immediately then drawn into a position between sides B(+) and C(0), in Chart XII.3B, (p.68), replacing the former dY1, down quark, [(PP/PP)(PP/VV)(PP/VV)], which we had shown to be redistributed into the newly formed up quark and antineutrino in the same chart.

The electron (e-) [(PP/PP)(PP/PP)(PP/PP)] and the antineutrino ($\overline{V}_E$) [(PP/PP)(PP/VV)(VV/VV)] were both ejected in this the reaction (Chart XII.3B).

The newly created up quark, a UR1r type, positioned between B(+) and C(0) in Chart XII.3B, together with the two quark remnants of the neutron shell in Chart XII.3, the dB1, down quark, and the other UR1r type up quark, now merge together to produce a proton shell (UUd), [(UR1r)(UR1r)(dB1)], shown in Chart XII.3B, (p.68) created from the original neutron shell [(UR1r)(dY1)(dB1)], in Chart XII.1.

The result of the neutron shell /g/ā meson decay process is thus its total conversion into a proton shell (UUd), an electron and an antineutrino (see Chart XII.3B, p.68).

The decay process also produces a complete conservation of photons: thirty V photons and thirty P photons at both the beginning and end of the process.

Chart XII.3A

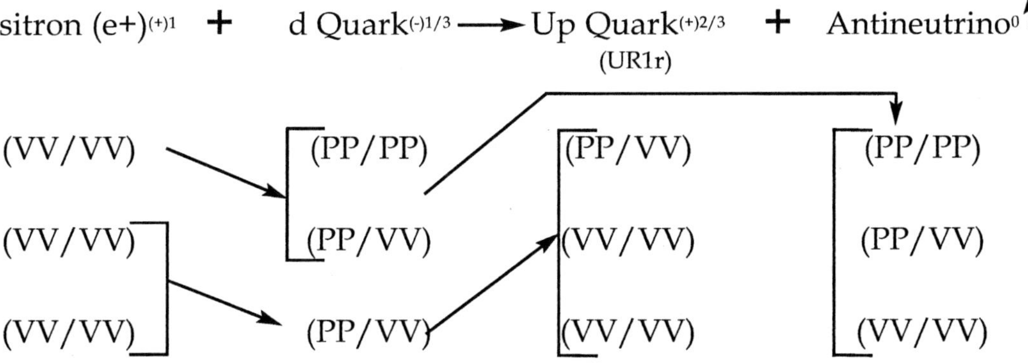

Chart XII.3B

A(0)
](+)1
PROTON
SHELL
(UUd)
[UR1r](+) 2/3
(+) (VV/VV) (VV/VV) (PP/VV) (+)
[dB1](-) 1/3
(-)(PP/PP) (PP/VV) (VV/PP)(-)
B(+)
(+) (VV/VV) (VV/VV) (PP/VV) (+)
[UR1r] (+) 2/3
C(0)

+
(PP/PP)
(PP/PP)
e-
(electron)
(PP/PP)
](-)1
ejected
+
(PP/PP)
(PP/VV)
$\overline{V}_E$
(antineutrino)
(VV/VV)
]0
ejected

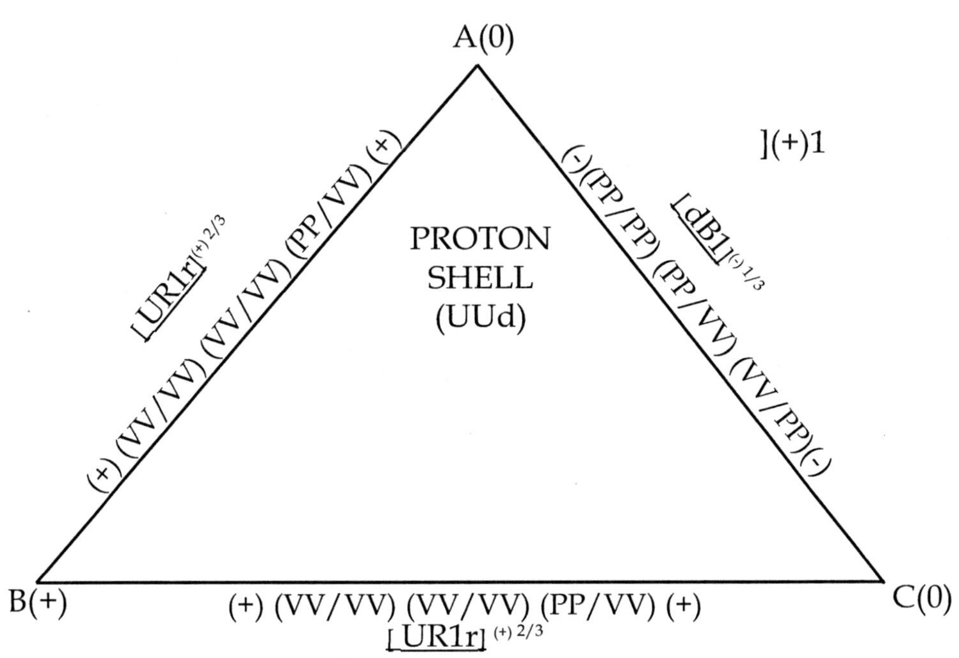

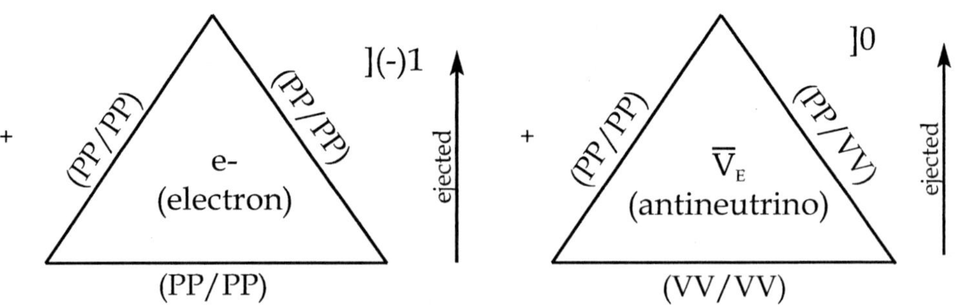

Proton and Neutron Shells

At the end of the proton decay process and the neutron decay process, the transformations have resulted in the production of a neutron shell (Udd) and a proton shell (UUd), respectively.

Both shell structures had originally formed around a central g/ā meson, which acted as a structural support to both. The meson as a nucleus in this context, also acts as would a wheel rim in a tire, greatly reinforcing and strengthening the proton or neutron shell surrounding it.

Once the meson has been removed due to the decay process, the now unsupported proton and neutron shells then behave like tires without a wheel rim, in that they are quite pliable, being able to bend and twist, especially in each of the three bond juncture points between ends of the three quarks in any proton or neutron shell present in any of the isotopes of the elements in the periodic table.

In the presence of a substantial energy flux, a proton or neutron shell therein can be twisted to the point that it breaks its bond to another proton or neutron which it is bonded to in a chain of such protons and neutrons in an element. This will result in the proton or neutron shell, and whatever is attached to it, separating from the main chain and becoming an independent, free unit. An alpha particle is an example of the end product of such a bond breaking. It will be discussed later in greater detail.

In summary, there are extant two types of protons: a proton shell /g/ā meson [(UUd/g/ā)$^{(+)1}$] and a proton shell [(UUd)$^{(+)1}$], both bearing a (+) 1 charge. There are, as well, two types of neutrons: a neutron shell / g/ā meson [(Udd/g/ā)^{0}] and a neutron shell [(Udd)^{0}], both exhibiting a zero charge.

All of the elements found in nature, or made in a laboratory, can contain mixtures of the two types of protons and two types of neutrons, except for hydrogen, which can be either type of proton.

Deuterium and tritium isotopes of hydrogen can also be composed of either type of protons: plus, in the case of deuterium, either one of the two types of neutrons; and in the case of tritium, two of the same type of neutron, or two different types of neutrons.

The proton shell (UUd) has twenty-four V photons, each at (+) 1/12 charge, and twelve P photons at (-) 1/12 charge. This is the equivalent of two positrons at (+) 1 and one electron at (-)1.

If we now add a g/ā meson to its center, which contains twelve V photons and twelve P photons, the proton shell /g/ā meson will contain thirty-six V photons and twenty-four P photons, the equivalent of three positrons and two electrons. The extra positron's twelve V photons will produce a (+) 1 charge on a proton shell /g/ā meson or a proton shell.

The two types of extant neutrons both bear a zero charge. The neutron shell (Udd) is composed of eighteen V photons and eighteen P photons.

When we add a g/ā meson into its center, producing a Udd/g/ā meson, it adds an additional twelve each of the V and P photons, bringing the V photon and the P photon numbers to thirty each, respectively.

Thus, the eighteen of each of the V and P photons found in the neutron shell would be the equivalent of one and one-half positrons and one and one-half electrons, resulting in a zero charge.

The thirty of each V and P photons found in the neutron shell /g/ā meson would equal two and one-half positrons and the same number of electrons, and a zero charge.

Along with the single form of the electron, the two types of protons and the two types of neutrons are the five bedrock entities from which all of our elements, planets and solar system are constructed.

But why cannot the (+) 1 positron, the zero charge neutrino and antineutrino also be used in the building of elements?

The positron is much too reactive, especially in respect to its interaction with the electron. It rips apart almost every particle it encounters, causing new entities to form. In many reactions, its monopoles or bosons are broken up and amalgamated with other monopoles or bosons from other interacting particles to produce new types of particles. Thus, it disappears as a specific particle by itself.

This reactivity is in great part due to its small size and homogenous twelve V photon content, which allows it to hit small sections of another entity, concentrating its (+) 1 punch there. The proton shell, on the other hand, contains three times as many total photons as the positron, while the proton shell /g/ā meson has five times the number as has the positron. The greater surface area size permits these protons, in contrast to the positron, to deliver the same (+) 1 charge to the same entity over a larger area when they are in contact. Thus, the contacted entities structural components are less apt to be ripped apart by the large area of this type of (+) 1 force than a small area by the small and concentrated positron's (+) 1 force.

The force contained in a large storm area may be equal to that contained in a tornado's small area; but the latter will rip everything apart in its contact area, while the former will only shake everything up, with moderate or little damage done.

The protons are huge in comparison to the positron, due primarily to the P photon content. The V photon size, in contrast to the P photon, is minute.

The V photon expresses its fundamental force as expansiveness. The P photon has no force, only a resistance to force. This resistance aids in stabilizing the proton relative to the positron, which has no P photons to stabilize it.

Precis : For pages 71 to 112

Four proton shell (UUd) reactions are diagrammed. In each reaction two possible reaction pathways can result; each pathway is composed of three specific reaction products. Each pathway is complex; and should be studied carefully.

Four neutron shell (Udd) reactions are shown to be reversible into the four proton shell reactants previously examined. In all of the cited reactions, the number and type of photons on the reactant side and on the reaction product side are exactly the same. All photons are thus conserved and accounted for on both sides.

Neutron decay processes, in one and two-step sequences, are presented in which a resultant proton shell is produced in a branch chain of an element, which then can be twisted to the point at which it's electric circuits are shut down momentarily, turning off it's (+)1 charge, which had allowed it to bond to another contiguous proton's (+)1 charge when both were in a formative magnetic dimension, where the rule is like charges attract and bond. After the twisting subsides, and its (+)1 charge is once more reinstated, the proton shell is now in an electric dimension in which like charges repel. Thus, it is now repelled by its formerly bonded to proton's (+)1 charge, instead of its being able to be bonded to it once more. Therefore, it and another contiguous proton shell, which it is bonded to, as well as two neutron shells still attached to both, are all moved away by the aforementioned repel action, resulting in their becoming a new four part entity, an alpha particle. Also, from the remainder of the element a new element emerges.

Antiprotons and antineutrons are herein diagrammed to show a method by which both are converted by interaction with a positron into a proton. Another possible outcome of these two antiparticles and a positron interacting is also diagrammed. Here, the outcome is the conversion of both antiparticles into: electrons, positrons, antineutrinos and neutrinos.

The neutron shell/g/ā meson also can react with a positron. Here, either another neutron shell/g/ā meson, or a proton shell g/ā meson is produced. Most neutron shell/g/ā mesons created, however, approximating one-quarter of all types of shell/g/ā mesons produced, are however, saved from conversion and disappearance by being reacted together with a proton shell/g/ā meson, thereby producing deuterium or tritium. The latter, in turn, and it's precious cargo of two neutron shell/g/ā mesons, is further saved from certain radioactive decay by being reacted with another proton shell/g/ā meson or deuterium, thus forming a stable isotope of the element helium, the second most numerous element in the solar system.

Four Additional Proton Reactions

We will examine the following four proton reactions:

1. Proton shell /g/ā meson (UUd/g/ā) plus an electron (e-), which interaction creates a neutron/ g/ā meson (Udd/g/ā) and a neutrino (V_E);
2. Proton shell (UUd) plus an electron (e-), which yields a neutron shell (Udd) and a neutrino (V_E);
3. Proton shell /g/ā meson (UUd/g/ā) plus an antineutrino ($\overline{V}_E$), which reaction creates a neutron shell /g/ā meson (Udd/g/ā) and a positron (e+);
4. Proton shell (UUd) plus an antineutrino ($\overline{V}_E$), which yields a neutron shell (Udd) and a positron (e+).

In the reactions in 1, 2, 3 and 4 above, only the proton shell is involved. Since the g/ā meson does not participate in the above interactions, the illustrations used to depict the reactions in 1 and 3 will also suffice for 2 and 4.

Proton and Electron Reaction

Chart XIII.1, p.73, depicts an electron (e-)$^{(-)1}$ and a proton shell /g/ā meson (UUd/g/ā)$^{(+)1}$ reacting with each other (+).

The electron is our standard three W- boson (PP/PP) type, but the proton shell structure with which it is reacting is a specific [UY1r][UY1][dY3] type. The g/ā meson therein is not further identified as to color, etc., since the g/ā meson does not partake in the reaction. Therefore, since the reaction is limited to the proton shell (UUd)$^{(+)1}$, it will be the same reaction for the proton shell /g/ā meson (UUd/g/ā)$^{(+)1}$, as well as for the proton shell entity itself.

The proton shell illustrated in Chart XIII.1A, p.73, is depicted as a triangle, with each of its three sides displaying a single quark: side AD is a dY3 down quark, while sides AP and DP are up quarks, [UY1r] and [UY1], respectively.

Each of the three quark sides has four letter points, which divide each quark into its three component bosons. For example, side AD is divided into three boson sections: AB (PP/VV), BC (PP/PP) and CD (PP/VV). The other two sides, AP and DP, are also divided into three sections, each one with a designated boson in it.

The two inner most letter points of the four letter points on each of the three sides of the outer triangle are then joined by drawing diagonal lines between them (see Chart XIII.2), (p.75). This produces two possible sets of three different inner triangles, seen in analysis of Charts XIII.2A and 2B, p.74 and on p.75, 2a and 2b.

The three inner triangles, Chart XIII.2A (Analysis) p.74, are one possible combination of three inner triangles; while Chart XIII.2B (Analysis), p.74, illustrates the second possible set and combination of three other inner triangles.

In nearly all reactions between one of the four principal structural entities (proton, neutron, antiproton and antineutron) and a small reacting entity (an electron, positron, neutrino or antineutrino), two possible sets of three different inner triangles can be produced; each of the inner triangles actually designating a specific new entity which has been formed in the reaction, due to a shifting of the bosons on the three sides of the principal structural entity.

An example of this type of reaction can be seen in Charts XIII.2C, (p.75) and 2D, (p.76). Here, the electron (e-), shown as a triangle with each side having a (PP/PP) boson, slams into the proton shell, triangle ADP, at section JP, which contains two (VV/VV) bosons, on its AP side (JH and HP).

Two sets of three possible entities and the reacting (/) electron (e-) are produced as a result. The first set, seen in Chart XIII.2C, p.75, consists of a d̄ quark, a d quark, a positron (e+) and the reacting (/) electron (e-); while the second set, in Chart XIII.2D p.76,

Chart XIII.1

Electron (e-)(-)1 +	Proton (UUd/g/ā)(+)1	⟶ Bottom	Top
(PP/PP)	Side AP	(VV/VV)(VV/VV)(VV/PP)	[UY1r]
(PP/PP)	Side DP	(PP/VV)(VV/VV)(VV/VV)	[UY1]
(PP/PP)	Side AD	(PP/VV)(PP/PP)(PP/VV)	[dY3]

Chart XIII.1A

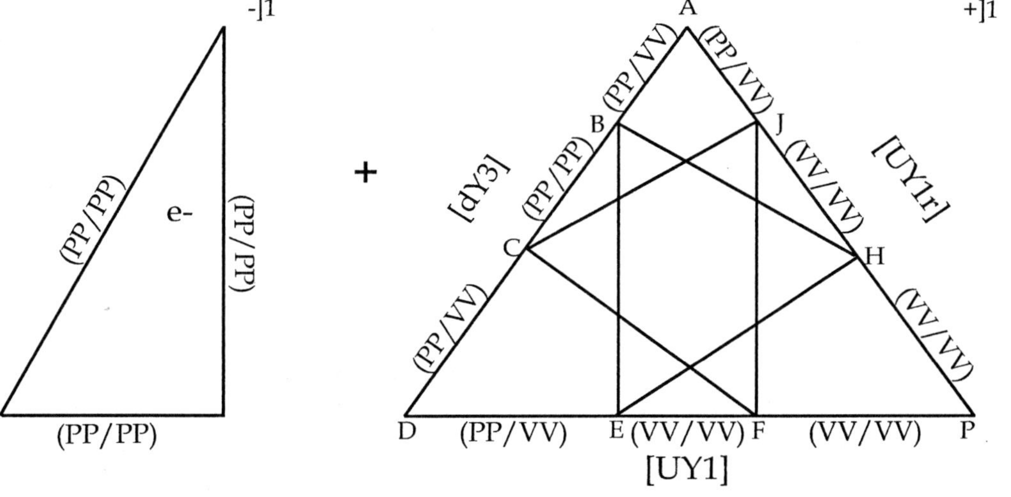

consists of exactly the same three particles, but in different positions.

In other types of reactions, however, we will see the formation of two distinctly different sets of three entities result, plus the intact reacting entity (/).

As soon as the three particles do form ($\bar{d}$, d, e+), in Chart XIII.2C, the electron (e-) and the positron (e+), △3 FPJ , will then react with each other, as depicted in Chart XIII.3A, p.77, forming a d quark and a $\bar{d}$ quark as a result.

The d and the $\bar{d}$ quarks then apprehend a third quark, the $\bar{d}$ quark (△1 CDF), which was contiguous to the positron and left unreacted, in Chart XIII.2C. This is necessary in order to attain a stable three quark configuration.

However, this three quark unit (d, $\bar{d}$, $\bar{d}$) is unstable, due to its net (+) 1/3 fractional charge, thus causing its constituent nine bosons to break their boson-to-boson bonds; then shift and rearrange themselves into an up quark (UR1r), a neutrino (V_E) and a down quark (dY1). This shift is depicted in Chart XIII.3B, (below),

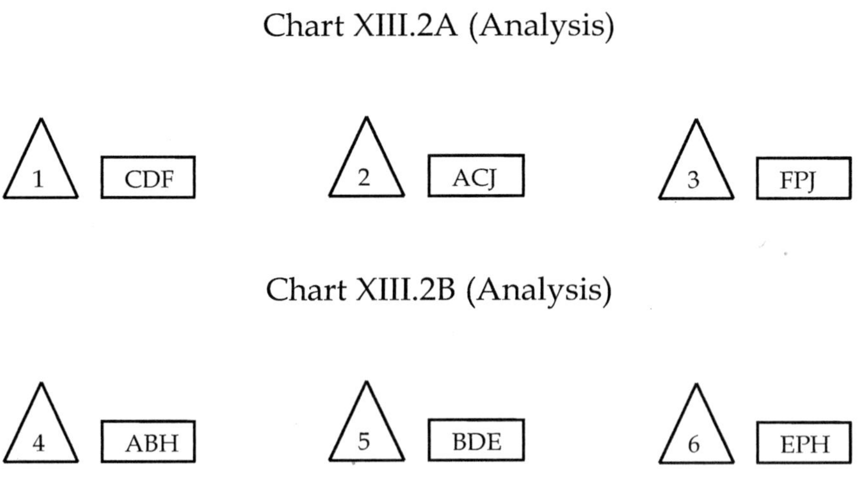

Charts XIII.2, 2A and 2B

2 | 2A | 2B

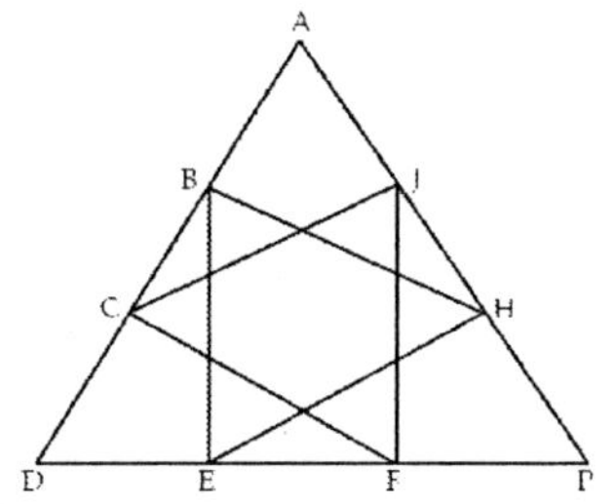

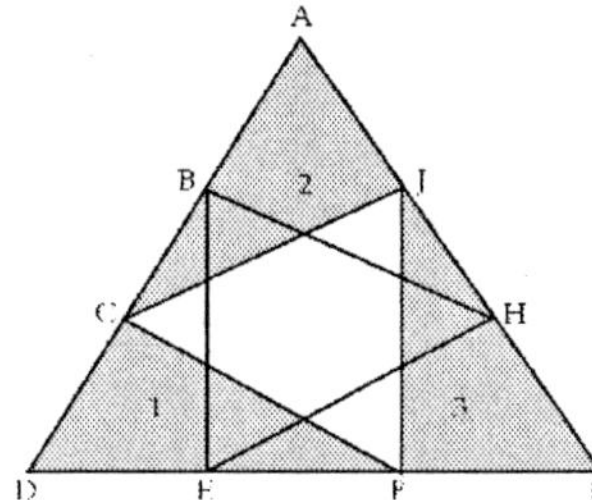

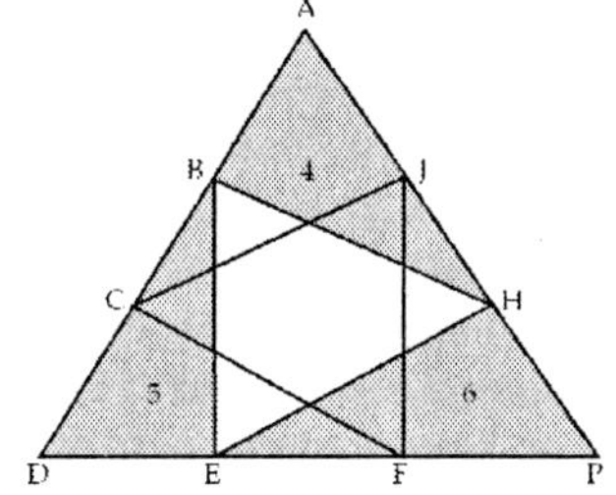

Chart XIII.2C

$\bar{d}$(△1 CDF)	+	d (△2 ACJ)	+	e+ (△3 FPJ)	+	(/) e-
(VV/VV)		(PP/PP)		(VV/VV)		(PP/PP)
(PP/VV)		(PP/VV)		(VV/VV)		(PP/PP)
(PP/VV)		(PP/VV)		(VV/VV)		(PP/PP)

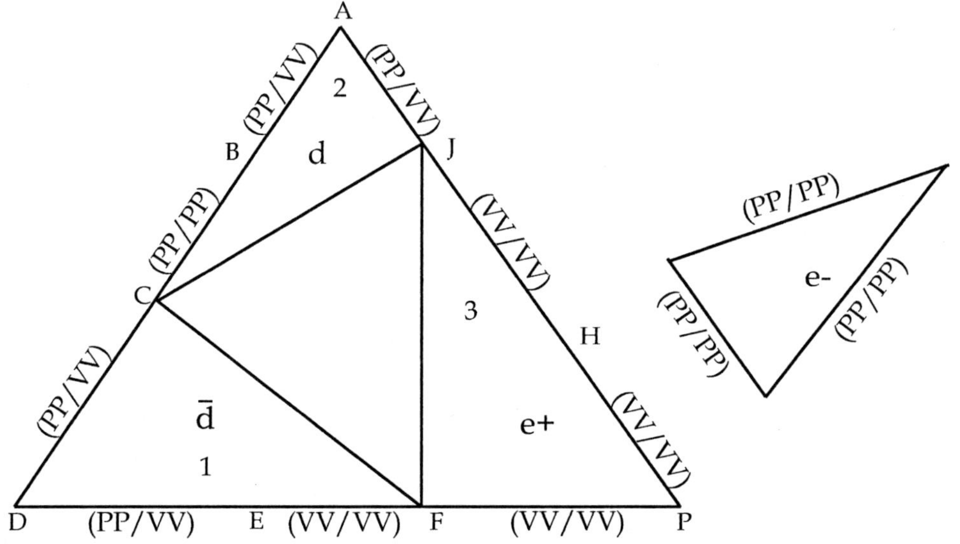

Chart XIII.2D

$\bar{d}$ (4 ABH)	+	d (5 BDE)	+	e+ (6 EPH)	+ (/)	e-
(VV/VV)		(PP/PP)		(VV/VV)		(PP/PP)
(PP/VV)		(PP/VV)		(VV/VV)		(PP/PP)
(PP/VV)		(PP/VV)		(VV/VV)		(PP/PP)

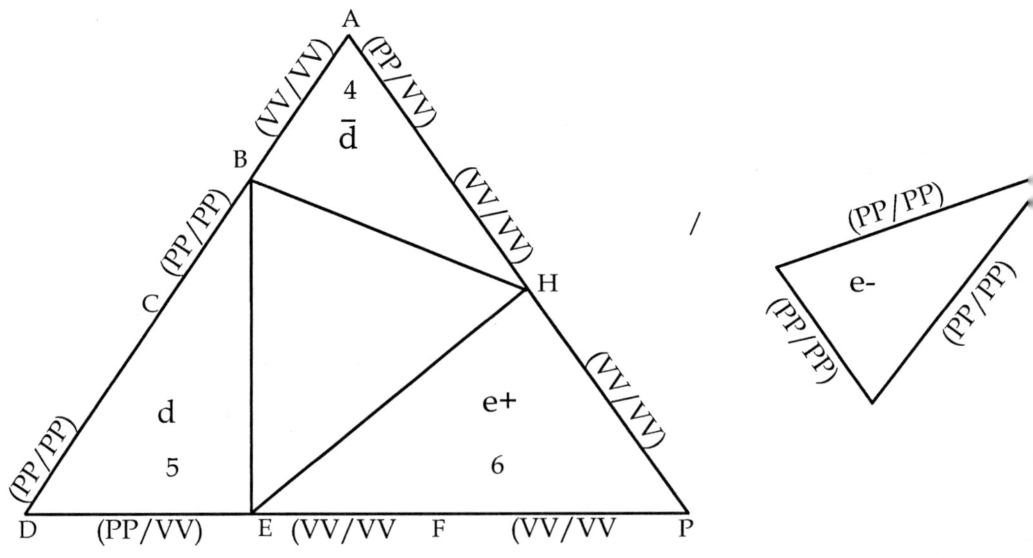

where rows 1, 3 and 2 reform into the respective UR1r, V_E and dY1 entities as a result of the respective free bosons formed from the former unstable, $\bar{d}$, and $\bar{d}$ quark unit structure.

The UR1r, dY1 and V_E unit thus formed is also unstable; which then results in the immediate ejection of the V_E (neutrino), leaving the remaining UR1r/dY1 complex to seek out a third stabilizing unit. The only one left is the d quark (△2ACJ), which is also classified as a dY1 type d quark, one of the unreacted, original three entities shown in Chart XIII.2C. It unites with the UR1r/dY1 complex to form a neutron shell $(UR1r/dY1/dY1)^0$ around the g/$\bar{a}$ nucleus, as shown in Chart XIII.3C, (p.78), bottom triangle.

It should be noted once more that the reaction shown in Chart XIII.3A, below, between the e- and e+ is quite violent, producing three free bosons in from each of these two reacting entities. It then

Chart XIII.3A

e-	+	e+	→	d	+	$\bar{d}$
(PP/PP)		(VV/VV)		(PP/PP)		(VV/VV)
(PP/PP)	⟷	(VV/VV)	→	(PP/VV)		(PP/VV)
(PP/PP)	⟷	(VV/VV)	→	(PP/VV)		(PP/VV)

Chart XIII.3B

	d	+	$\bar{d}$	+	$\bar{d}$ (△1CDF)	shift →	
Row 1.	(PP/VV)	+	(VV/VV)	+	(VV/VV)	→	UR1r
Row 2.	(PP/VV)	+	(PP/VV)	+	(PP/VV)	→	V_E
Row 3.	(PP/PP)	+	(PP/VV)	+	(PP/VV)	→	dY1

Chart XIII.3C

Neutron shell (Udd)				Neutron (V_E)
UR1r	+ dY1	+ d(2△ACJ)	+	V_E
(PP/VV)	(PP/PP)	(PP/PP)		(PP/VV)
(VV/VV)	(PP/VV)	(PP/VV)		(PP/VV)
(VV/VV)	(PP/VV)	(PP/VV)		(PP/VV)

↑ ejected

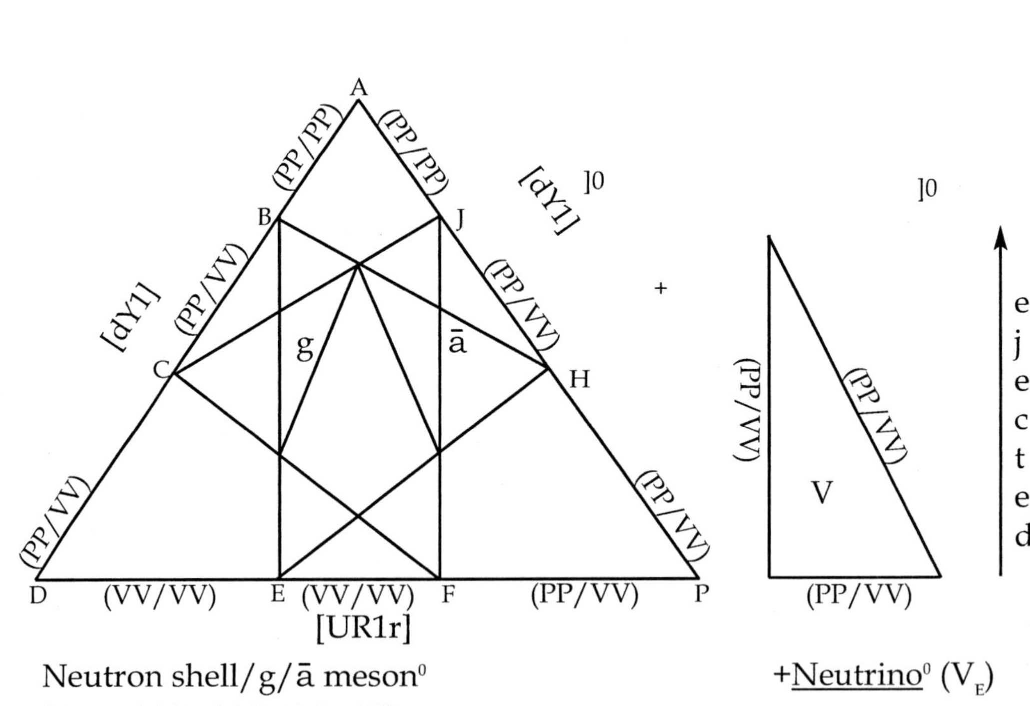

Neutron shell/g/ā meson^0 +Neutrino^0 (V_E)

[(UR1r)(dY1)(dY1)/g/ā]^0

[Udd/g/ā]^0

further causes two W-, (PP/PP) bosons from the e- and two W+, (VV/VV) bosons from the e+ to break their inner monopole-to-monopole bonds (/), thus resulting in the formation of four free (PP) and four free (VV) monopoles. These instantly reform as four Z bosons (PP/VV).

The two unreacted and intact bosons remaining from the e- and e+ reaction, a (PP/PP) and a (VV/VV), then each apprehend two of the four Z bosons (PP/VV), thereby creating the respective d and $\bar{d}$ quarks, seen in Chart XIII.3A, (p.77).

Proton and Antineutrino Reactions

In Chart XIV.1, (p.80), a proton shell /g/$\bar{a}$ meson$^{(+)1}$ (UUd/g/$\bar{a}$)$^{(+)1}$and an antineutrino ($\bar{V}_E$) are detailed as to their structure and composition, preliminary to entering into a reaction, (→).

Again, as in the previous proton and electron reaction, only the proton shell is involved. Thus, the reactions that will be illustrated are the same for both types of protons, (UUd/g/$\bar{a}^{(+)1}$) and (UUd$^{(+)1}$), even though just the proton shell is depicted in the reactions.

In Charts XIV.2A and 2B, (p.82 & 83, respectively) the antineutrino ($\bar{V}_E$) contacts (/) the proton shell on its AP side, at its HP section, containing a Z boson (VV/PP).

A reaction then ensues between the antineutrino ($\bar{V}_E$) and the proton shell, resulting in a freeing and shift of the nine bosons composing the three sides: AD, AP and DP, of the ADP triangle of the proton shell, into two different possible sets of three inner triangles (triads); each triad being a distinct type of entity composed of three of the nine newly free bosons. The two possible sets are shown in Charts XIV.2A and 2B, (p.82 & 83).

The first possible set of three triads which can result is: a $\bar{d}$ quark (▵HPE), a d quark (▵ABH) and a positron (e+) (▵BDE); while the ($\bar{V}_E$) is left unreacted and intact, (Chart XIV.2A), (p.82).

The second possible set of three triads is a ($\bar{V}_E$), a (V_E) and an (e+); which along with the still unreacted ($\bar{V}_E$) is seen in Chart

XIV.2B, (p.83). This second set of three entities is unstable, causing them to immediately seperate from each other and be ejected into the adjacent space, as is the unreacted $\overline{V}_E$, ending any further action as a unified entity.

In the first possible set: ($\bar{d}$, d, e+) and the contacting $\overline{V}_E$, the $\bar{d}$ and d quarks immediately form an unstable combination with the

Chart XIV.1			
Proton (UUd/g/ā)(+)1		+	Antineutrino $\overline{V}_E$ →
VV/VV)(VV/VV)(VV/PP	[UY1r]		(PP/PP)
VV/VV)(VV/VV)(PP/VV	[UR1r]		(VV/PP)
PP/VV)(VV/PP)(PP/PP	[dB1]		(VV/VV)

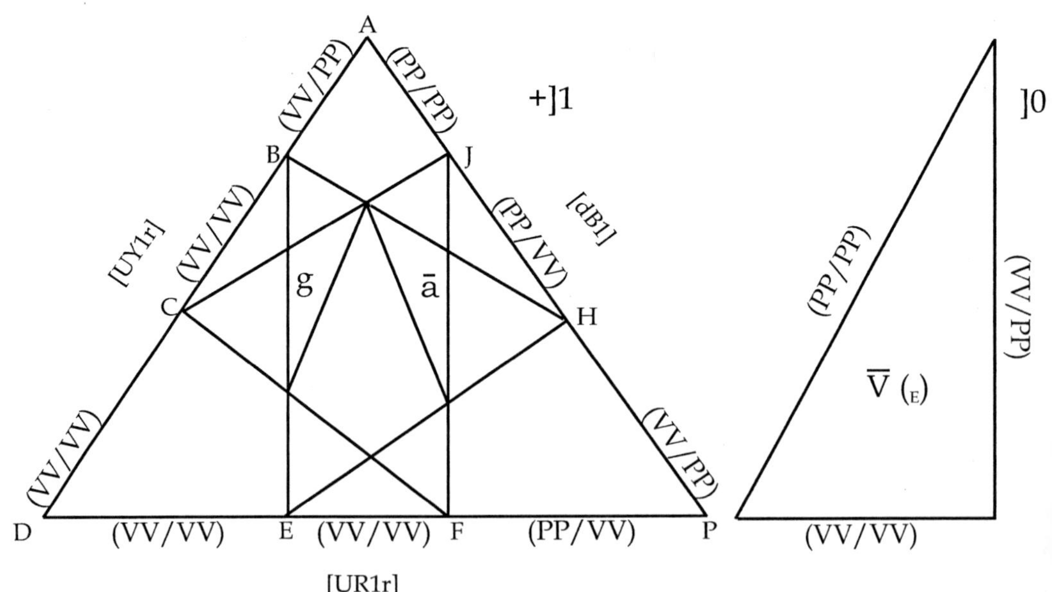

$\overline{V}_E$, as depicted in Chart XIV.3, (p.84). The e+ (ΔBDE) is ejected and is not shown in XIV.3.

Instability then ensues in the unstable combination above,, which causes a breaking of the boson-to-boson bonds present in all three ($\bar{d}$, d, $\overline{V}_E$), with a subsequent realigning of their bosons into three rows of quarks (Rows 1, 2 and 3, in Chart XIV.3, p.84).

Row 1 becomes a dB2r type of down quark [(PP/PP)(VV/PP)(VV/PP)]; Row 2 a dB1r, down quark [(VV/PP)(PP/VV)(PP/PP)]; while Row 3 results in a UR1r, up quark [(VV/VV)(VV/VV)(PP/VV)]. These three quarks then unite into a neutron shell $[(dB2r)(dB1r)(UR1r)]^0$, seen in the last column, Chart X1V.3 (p.84).

Four Neutron Reactions

We will now examine four different neutron reactions, as follows:

1. Neutron shell /g/ā meson plus a neutrino, which produces a proton shell /g/ā meson and an electron;
2. Neutron shell plus a neutrino, which yields a proton shell and an electron;
3. Neutron shell/g/ā meson plus a positron, which produces a proton shell/g/ā meson and an antineutrino;
4. Neutron shell plus a positron, which yields a proton shell and an antineutrino.

If we now compare these above four reactions with the four reactions illustrated and described in the section dealing with the "Four Proton Reactions" (p. 72), we will immediately see that the former are the exact reverse of the latter. For example, Number 1 above states that a neutron shell/g/ā meson and a neutrino produce a proton shell/g/ā meson and an electron; but Number 1 in the "Four Proton Reactions" shows that a proton shell/g/ā meson and an electron yield a neutron shell/g/ā meson and a neutrino.

Chart XIV.2A Set 1 Shifted Antineutrino/Proton Complex						
$\overline{V}_E$ (/)	+	$\bar{d}$ (△HPE)	+	d (△ABH)	+	e+ (△BDE)
(PP/PP)		(VV/PP)		(VV/PP)		(VV/VV)
(VV/PP)		(PP/VV)		(PP/PP)		(VV/VV)
(VV/VV)		(VV/VV)		(PP/VV)		(VV/VV)

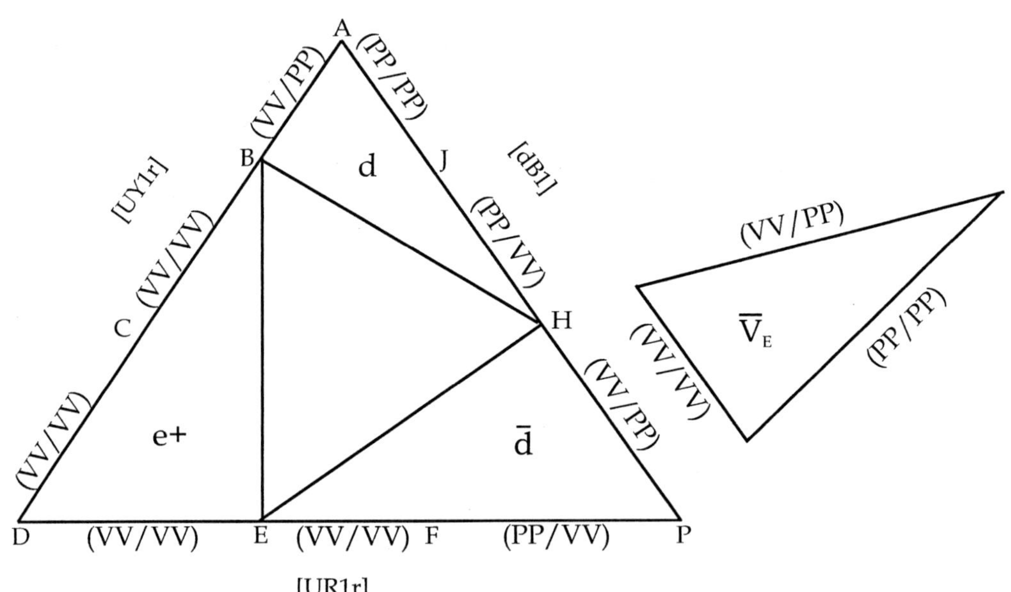

Chart XIV.2B

Set 2

Shifted Antineutrino/Proton Complex

$\overline{V}_{E-}$ (/)	+	V_E	+	e+	+	$\overline{V}_E$
(PP/PP) ↑ ejected		(PP/VV) ↑ ejected		(VV/VV) ↑ ejected		(PP/PP) ↑ ejected
(VV/PP)		(PP/VV)		(VV/VV)		(VV/PP)
(VV/VV)		(PP/VV)		(VV/VV)		(VV/VV)

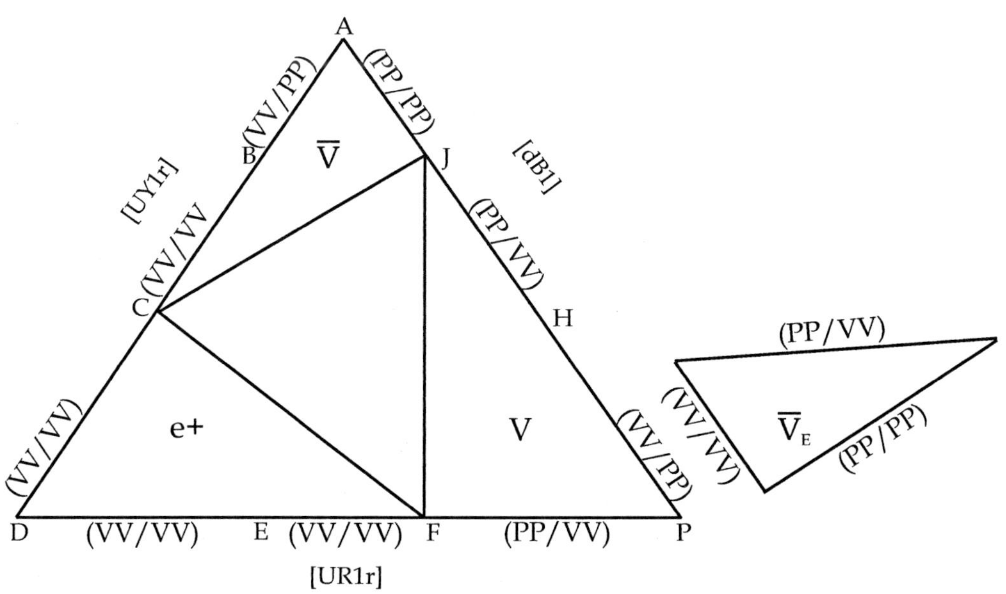

Chart XIV.3

$\overline{V}_E$	+	$\bar{d}$ (ΔHPE)	+	d (ΔABH)	⟶	Neutron Shell
Row 1. [(PP/PP)	+	(VV/PP)	+	(VV/PP)]	⟶	[dB2r]
Row 2. [(VV/PP)	+	(PP/VV)	+	(PP/PP)]	⟶	[dB1r]
Row 3. [(VV/VV)	+	(VV/VV)	+	(PP/VV)]	⟶	[UR1r]

All four neutron reactions and all four proton reactions are reversible reactions, which therefore should use the double arrows ⇄ instead of a single arrow, ⟶ to indicate reversibilty. For example:

$$(Udd/g/\bar{a})^0 + (V_E)^0 \rightleftarrows (UUd/g/\bar{a})^{(+)1} + (e-)^{(-)1}$$

The above neutron shell/g/ā meson $(Udd/g/\bar{a})^0$ and a neutrino $(V_E)^0$ react to form a proton shell/g/ā meson $(Udd/g/\bar{a})^{(+)1}$ and an electron $(e-)^{(-)1}$.

However, the above resultant electron $(e)^{(-)1}$ and proton shell/g/ā meson$^{(+)1}$ can then react to form the neutron shell/g/ā meson^0 and the neutrino $(V_E)^0$.

All of this rests on the assumption that in the case of the resultant proton shell/g/ā meson, the electron (e-), which also forms with it, does not sail off into space, thereby ending the reaction by making itself unavailable for further reaction.

In fact, in all of the previous cited eight reactions, the electron (e-), the antineutrino ($\overline{V}_E$), the neutrino (V_E) and the positron (e+) can effectively shut down any possible reverse reaction by leaving the reaction site by being ejected into a space beyond the field of reaction.

In all of the above cited reversible reactions, the number and the types of photons on both sides of the reaction are exactly the same. This is true for all reactions cited herein.

Single Neutron Decay in Elements

In an earlier exposition, we detailed the neutron decay process. An example of this process occurring in an element would be as follows:

$$^{24}_{11}\mathrm{Na} \rightarrow {}^{24}_{12}\mathrm{Mg} + (e\text{-})^{-1} + (\overline{V}_E)$$

The sodium atom, $^{24}_{11}$Na, contains eleven protons (the bottom number) and thirteen neutrons. This latter number is ascertained by subtracting the bottom proton number (11) from the top number (24), which is the addition product of the proton number (11) and the neutron number (13). This sodium atom is unstable and subject to radioactive decay or, specifically, neutron decay.

One of its thirteen neutrons, a neutron shell/g/ā meson type, due to an inability to form a magnetic bond between the A point on its neutron shell and the A' point on its g/ā meson, combined with an energy flux resulting from a previous neutron decay in a nearby area in the same sodium atom, initiates a neutron decay process sequence. The result is the creation of a proton shell $(UUd)^{(+)1}$, an electron $(e\text{-})^{(-)1}$, and an antineutrino $(\overline{V}_E)^0$, with the concomitant disappearance of the neutron shell/g/ā meson.

As a result, the protons in the sodium atom, $^{24}_{11}$Na, are now increased by the one additional proton shell. Therefore, we must raise the proton number to twelve (12). The proton number does not, however, distinguish between the number of proton shells and the number of proton shell/g/ā mesons present. A proton is a proton in counting the proton number.

With the loss of one neutron, replaced by the new proton shell, this raises the proton number to twelve and thus requires the element to lose its eleven proton number name, sodium, to be replaced by the signature name used for an element having twelve protons: $^{24}_{12}$Mg, magnesium.

The top number in both the $^{24}_{11}$Na and the $^{24}_{12}$Mg remains the same, due to having lost one neutron but having gained one proton, a zero net change.

The electron $(e)^{(-)1}$ and the antineutrino $(\overline{V})^0$ are also part of the signature of a neutron shell/g/ā meson decay process. Both of these are ejected from the immediate space of the decay transformation.

A proton shell, or a neutron shell, cannot undergo a respective proton or neutron decay process, since they do not possess a g/ā meson required in such decay processes. Only a proton/g/ā meson or a neutron/g/ā meson can therefore undergo a decay process.

Two Step Decay Process in An Element

A complex two-step series of reactions, which includes a neutron shell/g/ā meson decay process and the formation and ejection (↑) of an alpha particle, is seen below:

$$^{214}_{83}\text{Bi} \rightarrow (e-)^{(-)1}(\uparrow) + (\overline{V}_E)(\uparrow) + {}^{214}_{84}\text{Po} \rightarrow {}^{210}_{82}\text{Pb} + {}^{4}_{2}\text{He}^{+2}(\uparrow)$$

The atom of Bismuth, $^{214}_{83}Bi$, above, has one of its neutron shell/g/ā mesons undergo a neutron decay process, thereby producing both a signature electron $(e\text{-})^{(-)1}$ and antineutrino ($\overline{V}_E$), which are ejected into the adjoining space (↑), as well as a proton shell (UUd) which becomes a part of a newly formed element: Polonium, $^{214}_{84}Po$. As with the previously illustrated $^{24}_{11}Na$ (Sodium) into $^{24}_{12}Mg$ (Magnesium) process, the new proton shell created by the decay of a neutron shell/g/ā meson in the Bismuth, $^{214}_{83}Bi$, raises the proton number from eighty-three to eighty-four, thereby changing the signature name of the new eighty-four proton number element to Polonium, $^{214}_{84}Po$. The net change in this process (loss of a neutron, gain of a proton) is zero; therefore, the combined proton and neutron number of the Polonium atom remains at 214.

The newly created Polonium, $^{214}_{84}Po$, then undergoes a further transformation as a small chain of two protons and two neutrons breaks off one of the side branches coming off the main proton/neutron central spine of the element. The small chain is an ionized $^{4}_{2}$helium unit, $He^{(+)2}$; also called an alpha particle.

The above break is usually caused by an energy flux released by a neutron decay process in another nearby $^{214}_{83}Bi$ unit.

The alpha particle is ejected into the surrounding space. This leaves the $^{214}_{84}Po$ atom with eighty-two protons, the signature proton number of an atom of lead $^{210}_{82}Pb$. The number of neutrons present is one hundred and twenty-eight; plus the eighty-two protons, produces a top number of 210. The $^{210}_{82}Pb$ thus indicates the loss of the $^{4}_{2}He^{(+)2}$ unit from the $^{214}_{84}Po$.

But what has actually occurred? Let us first look at the composition and structure of the chain of protons and neutrons in the Bismuth, $^{214}_{83}Bi$, atom.

There is a central spine, composed of primarily proton shell/g/ā mesons and neutron shell/g/a mesons, which are bonded to one another. Coming off this central spine are branches, somewhat like a tree trunk spine and its leaf-bearing branches.

The branches, however, tend to have a majority of their component units as proton and neutron shells. These shell units are like tires without a wheel rim and are thus subject to breaking or tearing due to violent twisting caused by a nearby energy flux. The proton g/ā meson and the neutron g/ā meson, however, cannot be easily broken or torn due to the support of their g/ā meson nucleus, which acts as if it were a wheel rim.

The <u>strong force</u> is generated by all magnetic and electric charges present in the g/a meson, which, in specific cases previously illustrated, acts to bond or glue the g/a meson together, as well as bond the g/ā/meson to either a proton shell or a neutron shell; thereby creating a durable combination unit (shell g/ā/meson) which can then resist bond-breaking, thereby preventing a subsequent radioactive proton or neutron decay process (UUd/g/ā or Udd/g/ā). Also see p.183.It is especially effective where a magnetic bond is present between the A and Á charges, as previously shown in Chart IX, #7, p.48.

The <u>weak force</u> is generated by the magnetic and electric charges within a proton shell or a neutron shell, (UUd) or (Udd), and thereby bonds and glues each of these shell types together, (see p.184). It alone, however, is weak and can not prevent bond breaking and subsequent radioactive decay processes in a shell/g/ā meson.

On one of the branches in the $^{214}_{83}$Bi, there is a five-part terminal group, as seen in A. below:

A.

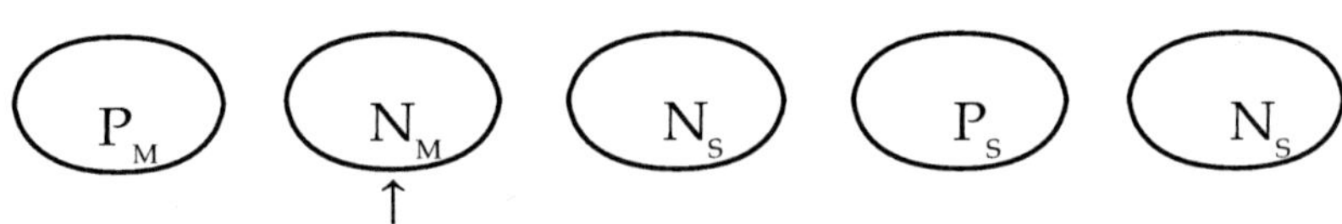

P_M = proton shell/g/ā meson
N_M = neutron shell/g/ā meson
N_S = neutron shell
P_S = proton shell

Initially, the N_M, above, with arrow ↑, undergoes a neutron decay process that converts it into a proton shell, P_S, plus an electron $(e-)^{(-)1}$ and an antineutrino ($\overline{V}$). The new atom that results from this neutron decay process is the Polonium, $^{214}_{84}Po$.

The five unit end group branch in A. above, in the $^{214}_{83}Bi$, has thus been transformed into branch B., below, still a branch chain off the central spine in the new $^{214}_{84}Po$. It now has a P_S, indicated by an arrow, ↑, which resulted from the decay of the N_M↑, in A. p.88.

B.

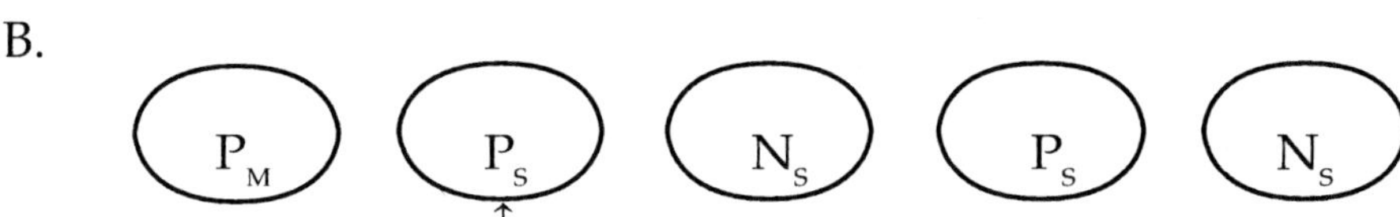

The P_S ↑, however, no longer contains a g/ā meson, which in a P_M or N_M acts as a stabilizing wheel rim so as not to permit any violent movement and subsequent bond-breaking in any of it's three junctions where the quarks are attached to one another.

The P_S ↑, without the g/ā meson, can have a bond break if a substantial enough energy flux is present; as the flux can shake and twist the P_S ↑ so violently as to cause the bond which joins one of its quarks to another quark in the attached P_M, at it's left side, to momentarily shut down their electric circuits and thus sever the bond between P_S↑ and the P_M. When the electric circuits once again reform, it will cause a now transformed (+)1 P_S, and the two N_S neutrons and the other P_S proton on its right side (the N_S, P_S, and $N_S)^{(+)1}$, now as a four-part unit, to be repelled and moved away from the now like charged (+)1 P_M, on it's left side, due to like charges repelling in this electric dimension (E11). Thus, it becomes an independent and free helium ion, $He^{(+)2}$, as shown in C. below:

C.

The P_M unit, to the left of the P_S unit, is still attached to the central spine of what is now an atom of lead, $^{210}_{82}Pb$.

Disappearance by Metamorphosis

The disappearance of the antiproton and antineutron from our universe has the makings of a great cosmic mystery story.

Like Monsieur Chevelan in the tale of the *Scarlet Pimpernel*, scientists have long sought these two elusive bodies, here, there and everywhere.

Only in the intimate confines of the laboratory have these wily creatures been captured for an instant.

We, however, know that both were created at the same time the protons and neutrons were, and each in quantities equal to those of the protons or the neutrons.

Theories projecting the existence of vast hidden galaxies, populated exclusively by antiprotons and antineutrons, waiting in cloaked ambush to interact with an ordinary proton and neutron galaxy, resulting in a great cosmic annihilation epoch, have been proposed.

The answer to their whereabouts, however, is that all the antiprotons and antineutrons in our universe have been transformed into protons, positrons (e+), electrons (e-), antineutrinos ($\overline{V}_E$) and neutrinos (V_E).

The reason for their metamorphosis rests entirely with their structural components and their properties.

Antiprotons

The three types of bosons, W+, W- and Z, compose the structure of the proton, neutron, antiproton and antineutron, which operate in an electric field in which the opposites attract and like repel rule is in effect.

The most powerful agent of change in an electric domain is the positron$^{(+)1}$ (e+), which is able to deliver a potent (+)1 charge, from its relatively small area volume to an equally small area volume in an antiproton$^{(-)1}$, antineutron^0 or neutron^0. Such a powerful impact on such a small area of any of these three large bodies will have the effect of shifting their bosons into two new types of arrangements, resulting in different quarks, neutrinos or other types of triads being created within these three bodies.

A proton$^{(+)1}$ is unaffected by a positron (e+) in an electric dimensional system, due to both having (+)1 charges which cause them to repel each other.

Additionally, the proton shell/g/ā meson$^{(+)1}$ (UUd/g/ā$^{(+)1}$) has a huge spatial volume of thirty-six V photons and twenty-four P photons, in comparison to the e+ which has only twelve V photons, and is thereby relatively tiny in comparison.

Thus, the proton delivers its potent (+)1 charge to another body over a relatively huge area, thereby limiting its impact and the disruption on the other body's boson components as it becomes contiguous; whereas, the e+ causes a huge impact with its (+)1 charge from its tiny body onto a tiny area of any entity with which it becomes contiguous, causing the bosons in that body to shift into either of two new configurations of different types of quarks, neutrinos, antineutrinos, electrons, etc.

The antiproton shell/g/ā meson$^{(-)1}$ ($\overline{UUd}$/g/ā$^{(-)1}$) has thirty-six P photons and twenty-four V photons, while the (e+) has twelve V photons. This can be confirmed by counting each type of photon in Chart XV.1, (p.92).

The former structure is depicted as an equilateral triangle herein, although its actual three-dimensional structure is far more complex. For teaching purposes, however, this triangular configuration will suffice.

Its $\overline{UUd}$/g/ā structure has two antiup quark sides, AD ($\overline{UY1}$) and DP ($\overline{UYI}$); and one antidown quark side, AP (dY1). (See Charts XV.1 and1A, p.92).

Its interior g/a meson is also shown as two contiguous triangles, a gY1r type and an āY 1 type.

Chart XV.1A shows the exact same $\overline{UUd}$/g/ā$^{(-)1}$ as Chart XV.1, but the interior diagonals used to form the two triangles have been removed so as to be better able to visualize the g/ā meson's component structures within the $\overline{Udd}$/g/ā$^{(-)1}$.

Chart XV.1
Antiproton Shell/g/ā Meson$^{(-)1}$ + Positron$^{(+)1}$ (e+)

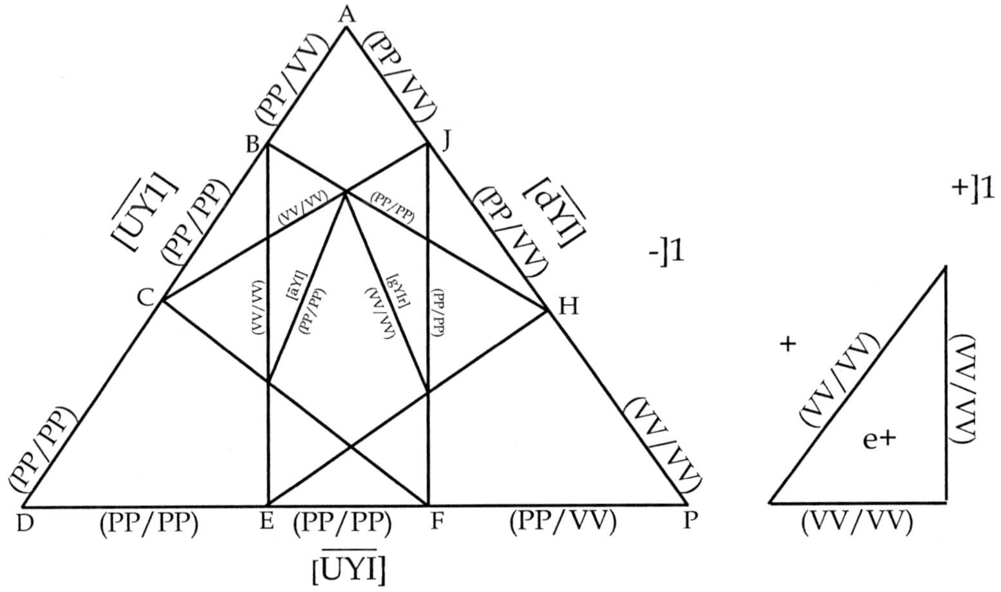

Chart XV.1A
Antiproton Shell/g/ā Meson$^{(-)1}$ + Positron$^{(+)1}$ (e+)

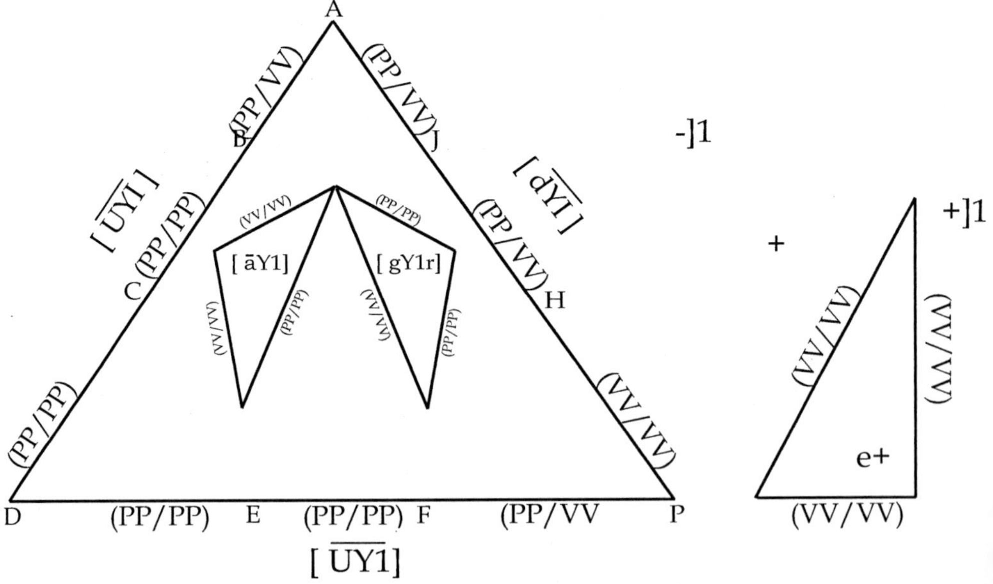

In both charts, each of the three sides of the $\overline{UUd}$ shell are further divided into three sections by the placement of two equidistant points from each end on each side (B, C; E, F; H, J).

Thus, side AD, representing a $\overline{UY1}$ type of antiup quark, shows its three boson components as AB (PP/VV), BC (PP/PP) and CD (PP/PP). Likewise, the same $\overline{UY1}$ antiup quark, on side DP, presents its bosons as sections DE (PP/PP), EF (PP/PP) and FP (PP/VV); while side AP exhibits its three sections, AJ (PP/VV), JH (PP/VV) and H (VV/VV), to be the boson components of its $\overline{dY1}$, antidown quark.

The g/ā meson, in the center of the $\overline{UUd}$ shell, manifests as two contiguous triangles: the gY1r gluon [(VV/VV)(PP/PP)(PP/PP)] and the āY1 antigluon [(PP/PP)(VV/VV)(VV/VV)].

These two triangles, ▵ gY1r and ▵ āY1 , are formed in Chart XV.1. (p.92), by drawing diagonal lines from each of the inner two equidistant points on each side to one of the equidistant points on each of two opposite sides. These six diagonal lines then outline a hexagonal interior structure, in which two smaller diagonal lines are drawn to form two small triangles, representing the gluon (▵gY1r) and antigluon (▵āY1) components of the meson. In Chart XV.1A, (p.92), the parts of the diagonal lines not outlining the two triangles are removed to better view the meson relative to the $\overline{UUd}$ shell structure surrounding it.

We will now turn to the two reaction pathways which can occur when the $\overline{UUd}$/g/$\bar{a}^{(-)1}$ and the e+, depicted in Chart XV.1A, become contiguous. The first reaction pathway is seen in Chart XV.2, (p.94). Here, the e+ strikes the W+ boson (VV/VV), section HP, of the antidown quark ($\overline{dY1}$), side AP, depicted in the diagram at the bottom of Chart XV.2.

This causes the nine bosons in the three antiquarks in the $\overline{UUd}$ shell to have their boson-to-boson bonds broken, thereby freeing all of them; which allows them to then reconfigure into the first of the two new sets of three boson reaction products, as seen in Chart XV.2, (p.94), the bottom diagram: an e- (▵CDF), a d quark (▵ACJ) and a $\bar{d}$ quark (▵JFP). The e+, which caused this reaction and reconfiguration, remains intact, and is also shown in Chart XV.2 "Reaction Products".

Chart XV.2

REACTION:

[Antiproton Shell/g/ā Meson$^{(-)1}$ ($\overline{UU}$d /g/ā$^{(-)1)}$) + Positron$^{(+)1}$ (e+)]

(yields)

REACTION PRODUCTS

e-(△CDF)	+	d(△ACJ)	+	d̄(△JPF)	+	e+
(PP/PP)		(PP/VV)		(PP/VV)		(VV/VV)
(PP/PP)		(PP/VV)		(VV/VV)		(VV/VV)
(PP/PP)		(PP/PP)		(PP/VV)		(VV/VV)

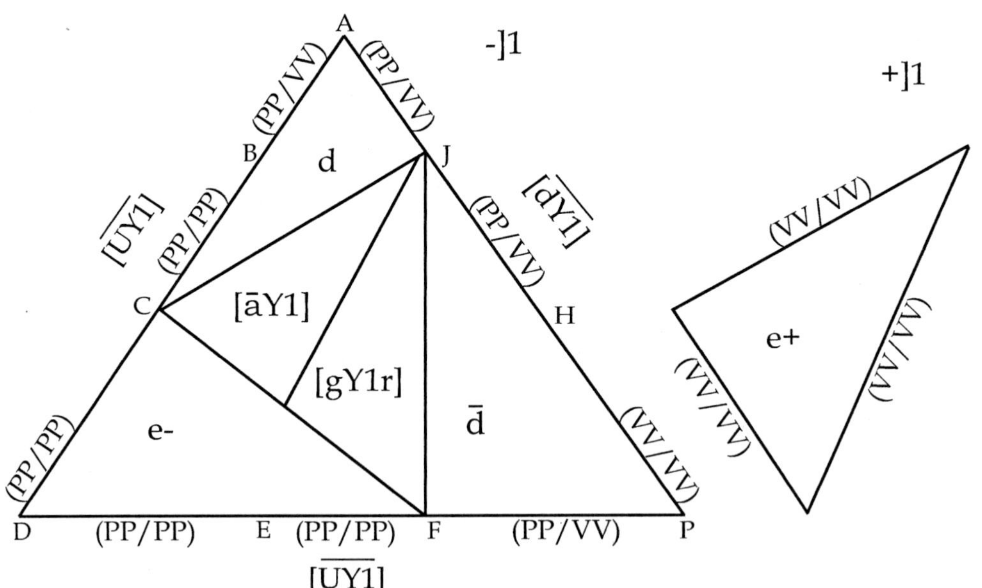

The next step in this first reaction pathway is shown in Chart XV.3, below, as the initiating e+ and the d̄ quark (▵ JFP), which was formed by the (e+) at the site of it's initial contact with the W+ boson (VV/VV), then further react with each other to yield (→) two new up quarks: a UR1r and a UR2 type.

The remaining unreacted d quark in the reaction products, (▵ACJ), a dY1r type down quark, p.94, then attaches itself to these two up quarks, (Chart XV.4), p.96, producing a new proton shell (UR1r, UR2, dY1r), which now surrounds the original, unreacted g/ā meson at its center. This newly formed proton shell (UR1r, UR2, dY1r)/g/ā (gY1r, aY1) g/ā/meson$^{(+)1}$ is depicted in Chart XV.5 p.96; as is the electron, e- (▵CDF), which was formed in the initial reaction with the e+ in Chart XV.2, p.94, which now is ejected into the surrounding area; or may neutralize the proton shell/g/ā meson$^{(+)1}$ just produced, forming a hydrogen atom.

Thus, the antiproton shell/g/ā meson$^{(-)1}$ has been converted into a proton shell/g/ā meson$^{(+)1}$ and an electron$^{(-)1}$ (e-).

This forgoing reaction, its reaction products and the two final products of the entire process just described are, however, only one of two reaction pathways that could have been followed in the reaction between the antiproton shell/g/ā meson$^{(-)1}$ and the positron (e+).

Chart XV.3

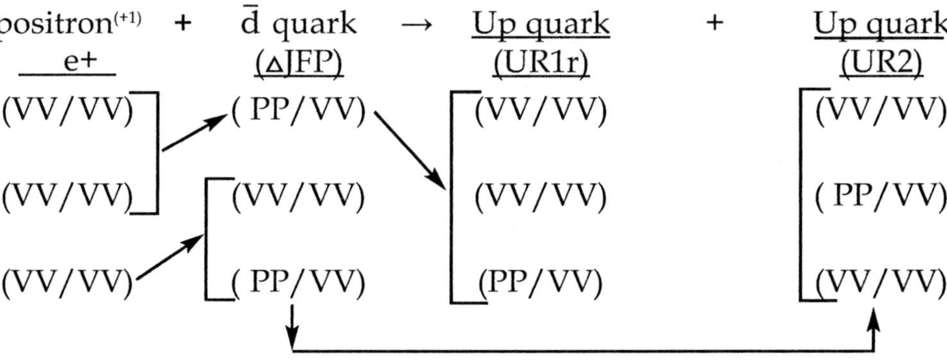

Chart XV.4
Proton Shell Formation

Proton Shell				+ electron	
UR1r	+ UR2	+ d(ΔACJ)	(+)1	e-	ejected ↑
(VV/VV)	(VV/VV)	(PP/VV)		(PP/PP)	
(VV/VV)	(PP/VV)	(PP/VV)		(PP/PP)	
(PP/VV)	(VV/VV)	(PP/PP)		(PP/PP)	

Chart XV.5

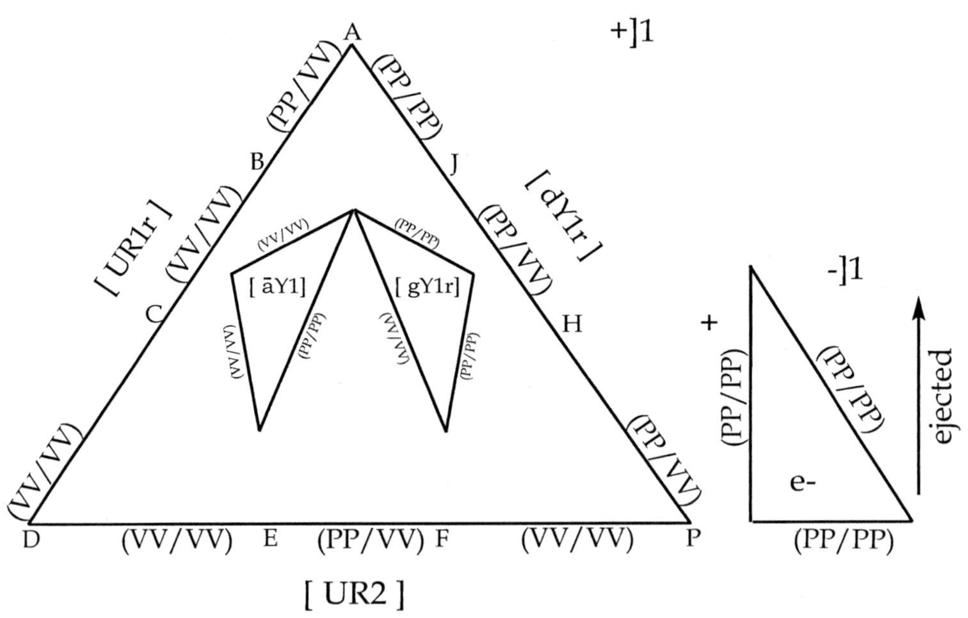

The second reaction pathway also begins with the e+ impacting the antiproton shell/g/ā meson$^{(-)1}$.

Chart XV.6A, (p.98), graphically depicts this. Here, however, the e+ strikes the AD side at the CD section, containing the W- boson (PP/PP), causing the antiproton shell to reconfigure into an e- (electron$^{(-)1}$), V_E (neutrino), and $\overline{V}_E$ (antineutrino), as well as also forcing the conversion of the g/ā meson into an e+ (positron$^{(+)1}$) and an e- (electron$^{(-1)}$), while the impacting positron$^{(+)1}$ (e+) still remains intact.

Chart XV.6B, (p.98) depicts these six products of this second reaction pathway: two e-, two e+ (including the impacting e+), one $\overline{V}_E$ and one V_E, all of which are ejected into the surrounding area as separate units.

If we contrast these six end products of pathway two, above, with those of pathway one, previously shown as a proton shell/g/ā meson$^{(+)1}$ and an electron (e-), we will note an important initial difference between the two pathways.

In Chart XV.2, (p.94), in the diagram at the bottom, the e+ strikes the antiproton shell on the AP side at section HP, where a W+ boson (VV/VV) is present; while in Chart XV.6A, (p.98), the hit is made by the e+ on the AD side, section CD, in which a W- boson (PP/PP) is in place.

In the former reaction (Chart XV.2, p.94) the reconfiguration produces a $\overline{d}$ quark at the point of impact, which then reacts with the impacting e+ to create two up quarks (Chart XV.3), (p.95). The intensity of this reaction, however, is not sufficient to alter the g/ā meson, and it remains intact.

In the latter reaction (Chart XV.6A, p.98) the e-, produced by the e+ at the point of impact, then immediately reacts with the same e+, in the most intense interaction between any two triads, thus releasing a great energy flux, causing the conversion of the adjoining g/ā meson into an e+ and an e-, as well as breaking of all the bonds holding together all of the newly formed five inner units (e-, e+, e-, V_E and $\overline{V}_E$).

These five inner units, which previously made up the antiproton shell/g/ā meson$^{(-)1}$, together with the impacting e+, then separate and are ejected into the adjoining area as six independent, free units.

As a result of this second type route of reaction, two e+ are now available for future reactions with other antiproton shell/g/ā mesons, etc., instead of just the one impacting, e+ unit.

Thus, this reaction route will effectively add additional positrons (e+) necessary for the swift and complete conversion of all of the $\overline{UUd}$ /g/$\bar{a}^{(-)1}$ units (antiprotons) extant into UUd/g/$\bar{a}^{(+)1}$ units (proton/g/ā meson$^{(+)1}$), as well as e-, V and $\overline{V}_E$.

Chart XV.6A

Positron$^{(+)1}$ (e+) + Antiproton Shell/g/ā Meson$^{(-)1}$

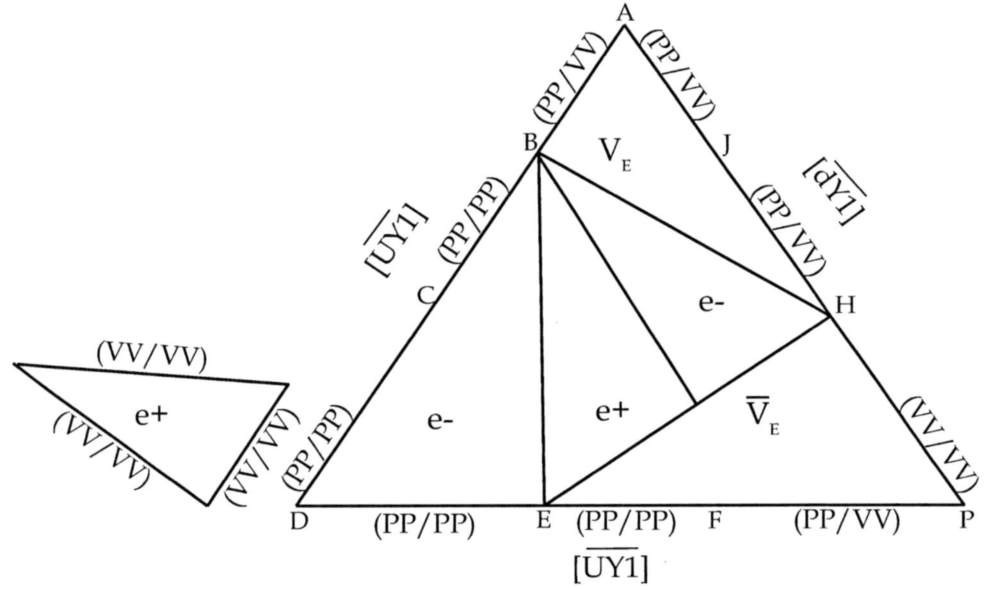

Chart XV.6B

Reaction Products

Number	Type	Symbol
2	Electrons	(e-) ↑
2	Positrons	(e+) ↑
1	Antineutrino	($\overline{V}$) ↑
1	Neutrino	(V) ↑

Chart XVI.1

Positron (e+) + Antineutron Shell /g/ ā Meson^0

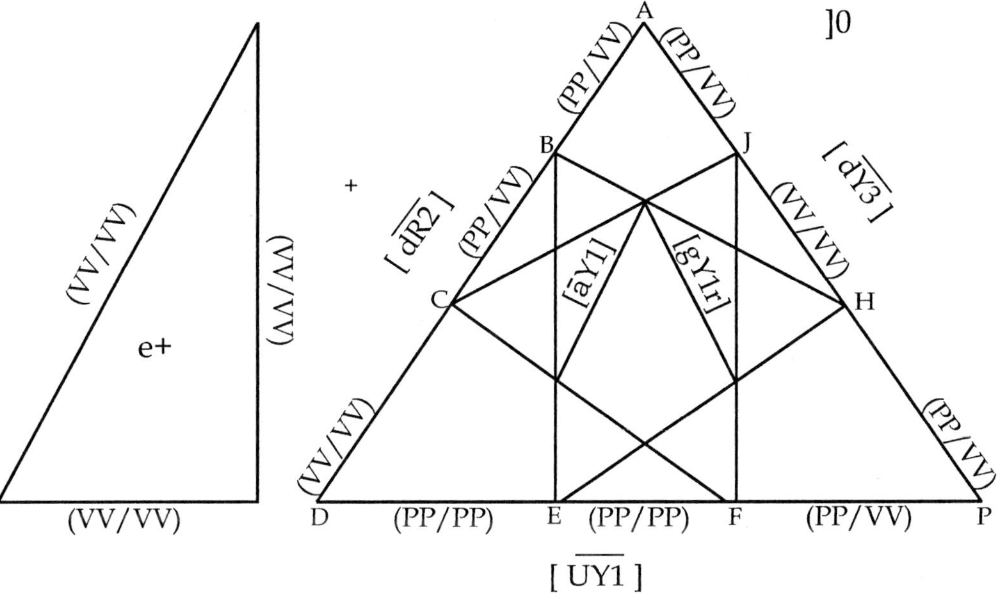

Antineutrons

The actions of the antineutron shell/g/ā meson^0 parallels that of the antiproton shell/g/ā meson$^{(+)1}$.

An e+ impacting the former will also cause either of two types of reconfigurations of its three sided antiquark shell structures and g/ā meson.

Chart XVI.1, above, depicts an antineutron shell/g/ā meson^0 ($\overline{U}d\overline{d}$/g/ā)^0: $\overline{UY1}$, $\overline{dR2}$, $\overline{dY3}$ /gY1r / āY1, and a positron$^{(+)1}$ (e+), which will react with it.

Charts XVI.2A and 2B, (p.100) illustrate the two possible reconfigurations, or routes, which the $\overline{U}d\overline{d}$/g/ā in Chart XVI.1 can produce as the result of its interaction with an e+.

Chart XVI.2A, (p.100), depicts the first reaction route, which ends with the formation of two quarks (d, $\overline{d}$) and an antineutrino ($\overline{V}_E$). This type of unit will further react.

Chart XVI.2A
Positron/Antineutron Shell /g/ā Meson Complex (First Reaction Pathway)

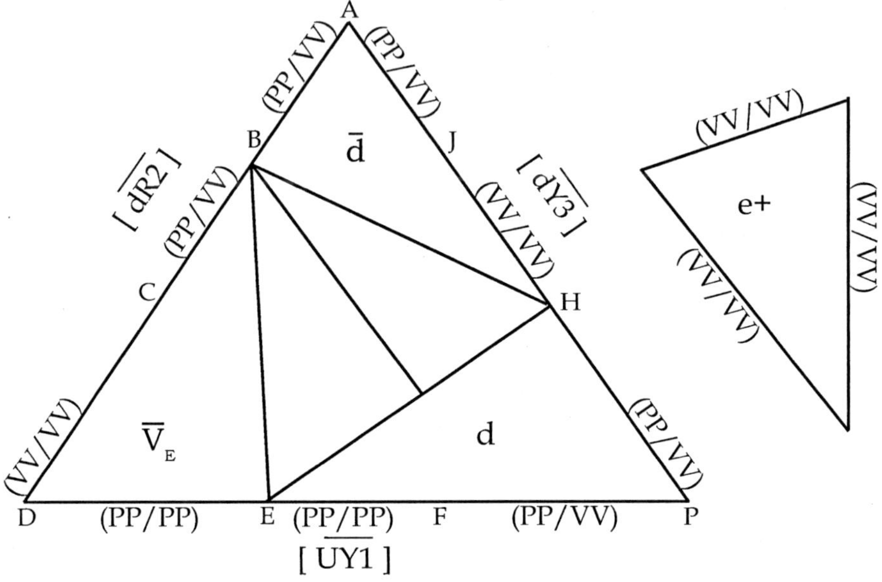

Chart XVI.2B
Positron/Antineutron Shell /g/ā Meson Complex (Second Reaction Pathway)

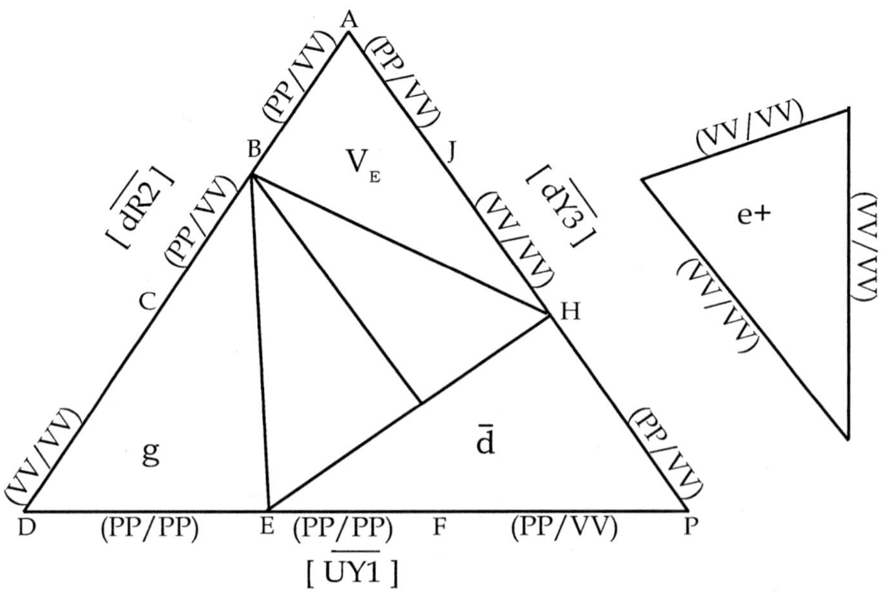

Chart XVI.2B, P.100, depicts the results of the second reaction pathway, which produces a single gluon (g), along with a $\bar{d}$ and V_E.

A lone gluon cannot be formed by itself. It must always be bound to an antigluon in it's formation. This is not a matter of stability but a basic understanding that a gluon/antigluon meson is really a positron[(+)1] (e+) and an electron[(-)1] (e-) in a tenuous equilibrium, in which a single W+ boson (VV/VV) and a single W- boson (VV/VV) have been exchanged by the two.

Thus, the structure in Chart XVI.2B cannot form, or is a forbidden sequence.

Therefore, in the reaction in Chart XV.1, only one possible reconfiguration pathway, shown in Chart XVI.2A, p.100, can be produced by the e+ impact.

Chart XVI.3

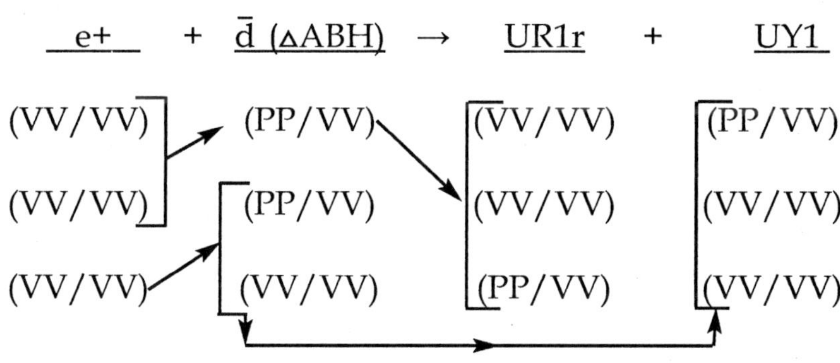

Chart XVI.4

Proton Shell / g/ā meson					+	Antineutrino
UR1r	+	UY1	+	d(△HEP)[dY1r]		$\bar{V}_E$
(VV/VV)		(VV/VV)		(PP/VV)		(PP/PP)
(VV/VV)		(PP/VV)		(PP/VV)		(VV/VV)
(PP/VV)		(VV/VV)		(PP/PP)		(PP/VV)

ejected

Chart XVI.5

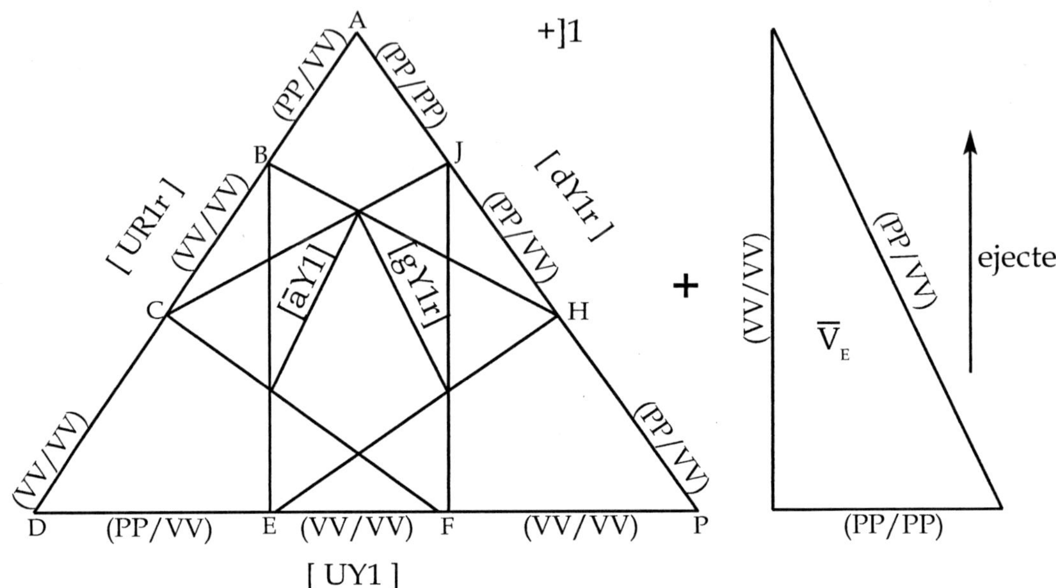

As a result of the reaction, which is seen in Chart XVI.2A, p.100, the newly created $\bar{d}$ quark ($\triangle$ABH) then reacts with the impacting e+ to produce two up quarks: a UR1r and a UY1 type, as depicted in Chart XVI.3, p.101.

The mechanism for this to take place is based on the e+ impact causing its three same W+ bosons (VV/VV) and the three bosons of the $\bar{d}$ quark being hit, to have all of their boson-to-boson bonds broken simultaneously; thereby creating six free bosons. These then bond as follows: Two W+ bosons (VV/VV) from the e+ and a Z boson (PP/VV) from the $\bar{d}$ quark combine to form a UR1r, up quark; while the same combination of two W+ bosons (VV/VV) and a Z boson (PP/VV) bond together in a slightly different order to produce a second up quark, a UY1 type. On both sides of the reaction (→) there are four W+ bosons (VV/VV) and two Z bosons (PP/VV). The reaction components on both sides are thus conserved.

The two newly synthesized up quarks then attract the unreacted d quark ($\triangle$HEP), a dY1r type, from Chart XVI.2A, p.100, and all three then bond together around the unreacted g/ā meson, thereby producing a proton shell (UR1r, UY1, dY1r)/g/ā meson (gY1r/āY1). This is seen in Chart XVI.4, p.101; together with the last remaining structure from the impact in Chart XVI.2A, p.100, the antineutrino ($\overline{V}_E$), which was simultaneously ejected.

The foregoing series of reactions is summarized and illustrated in Charts XVI.4 and 5, (p.101 and p.102).

Chart XVII.1, p.104, depicts another antineutron shell/g/ā meson type structure. Here, a dY1 antidown quark has been substituted for a dY3 antidown quark in the antineutron shell/g/ā meson, as illustrated in Chart XVI.1, p.99. Other than this substitution, both structures are identical.

The impact of an e+ on the antineutron/g/ā meson can result in two possible reconfigured structures or pathways. (See Charts XVII.2A and 2B, p.105)

In Chart XVII.2A, one of the three possible triads which may result is the g (gluon). This structure is forbidden, for reasons discussed previously. Thus, the entire structure will not form.

The structure in Chart XVII.2B however, is shown to have reconfigured into two $\overline{V}_E$ and one V_E, as well as an e+ and an e- from a reconfiguration of the g/ā meson.

These five units and the impacting e+ then separate from each other and are all ejected (↑) into the surrounding area. (See Chart XVII.3, p.106.)

Here, we see that two e+ units are now available at the end of this series of reactions, instead of one e+ which initiated the reaction.

This is a positron producing chain reaction; foretelling the parallel neutron generating chain reaction in elements such as uranium.

Chart XVII.1
Antineutron (Udd) + Positron (e+)

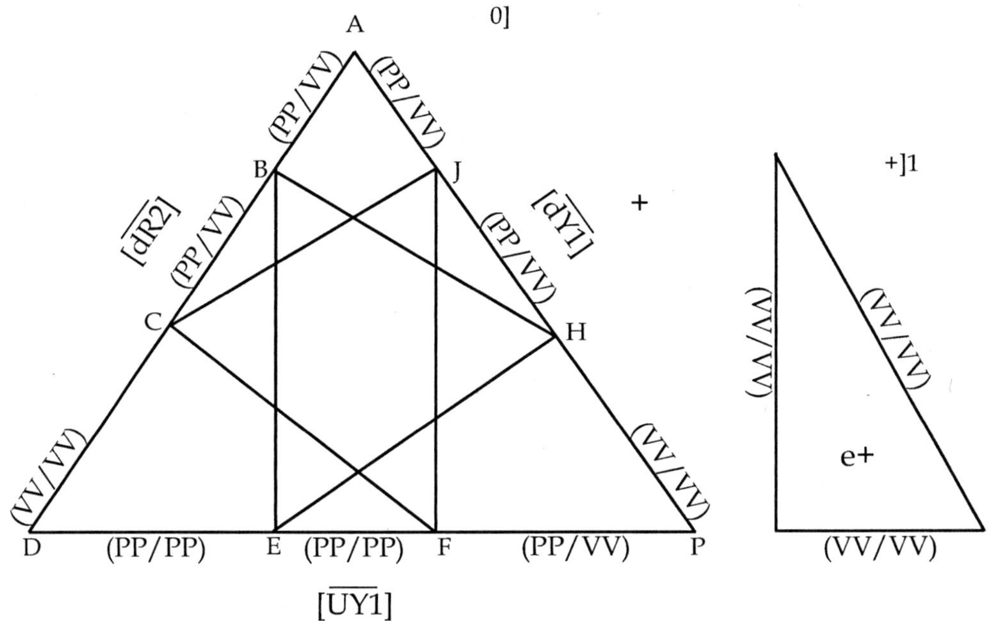

Chart XVII.2A
(First Reaction Pathway)

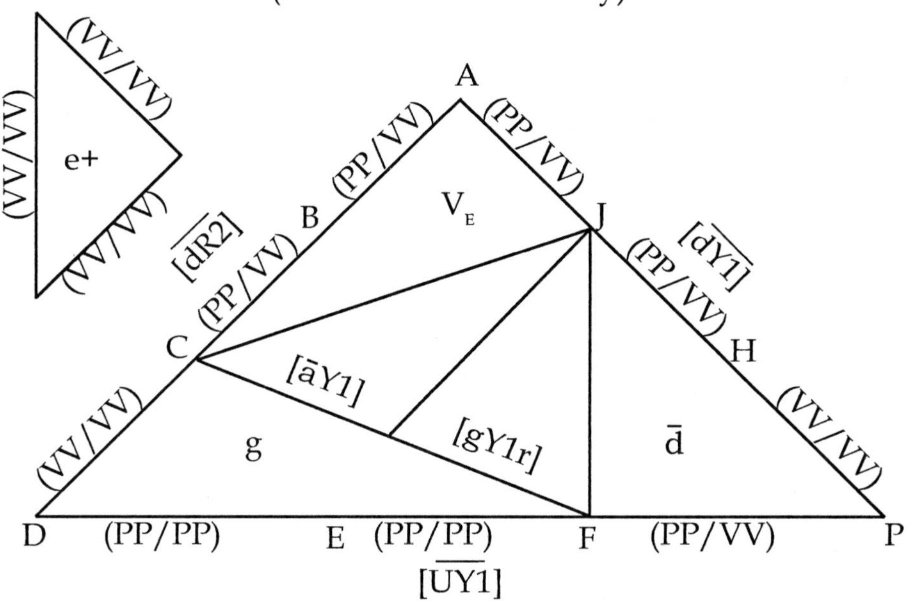

Chart XVII.2B
(Second Reaction Pathway)

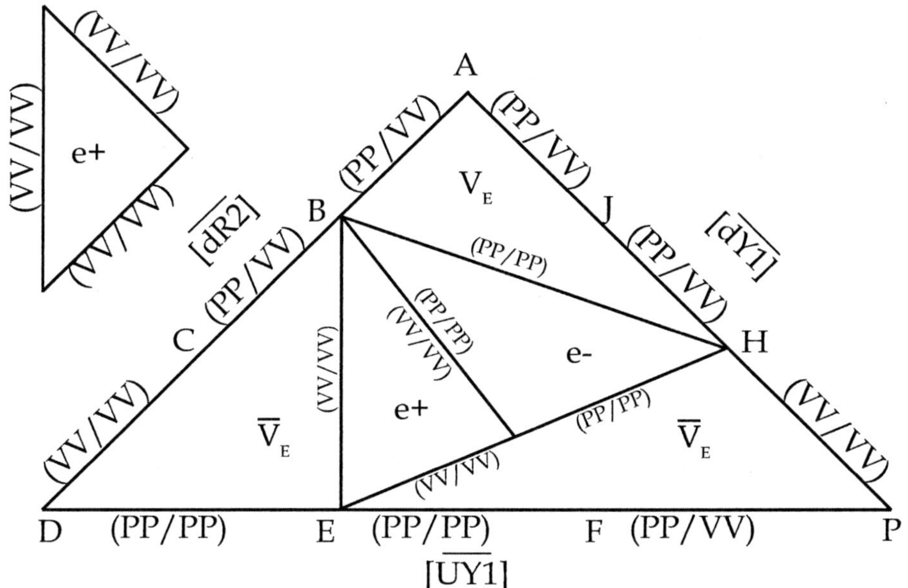

Chart XVII.3

Ejected Components

Number	Type	Symbol	
1	Electron	e-	↑
2	Positrons	e+	↑
2	Antineutrinos	$\overline{V}_E$	↑
1	Neutrino	V_E	↑

With the doubling of available positrons (e+), the conversion of all antineutron shell/g/ā mesons is accelerated. Eventually, all such $\overline{Udd}$ /g/$\bar{a}^0$ units disappear from the existent universe, replaced by proton shell/g/ā mesons, V_E, $\overline{V}_E$, e- and e+ units.

Neutrons

A neutron shell/g/ā meson^0, Udd/g/$\bar{a}^0$, parallels the antiproton shell/g/ā meson$^{(-)1}$, $\overline{UU}$d/g/$\bar{a}^{(-1)}$, and the antineutron shell/g/ā meson^0, $\overline{UU}$d/g/$\bar{a}^0$ meson, in its reaction with a positron$^{(+)1}$, e+. All three of these structures can form two possible reconfiguration pathways of their original all quark structure as a result of the e+ interaction.

Chart XVIII.1, (P.107), depicts one type of Udd/g/$\bar{a}^0$ (neutron^0) about to be impacted by an e+.

Charts XVIII.2A and 2B, (P.108), illustrates the two types of reconfigured structures which can result from the e+ impact.

Chart XVIII.2A reveals that this first type of reconfigured structure is just another type of Udd/g/$\bar{a}^0$. It will remain intact unless impacted again by another e+ or other triad capable of effectuating a modification.

Chart XVIII.1
Neutron Shell/g/ā Meson⁰ + Positron(+)1(e+)

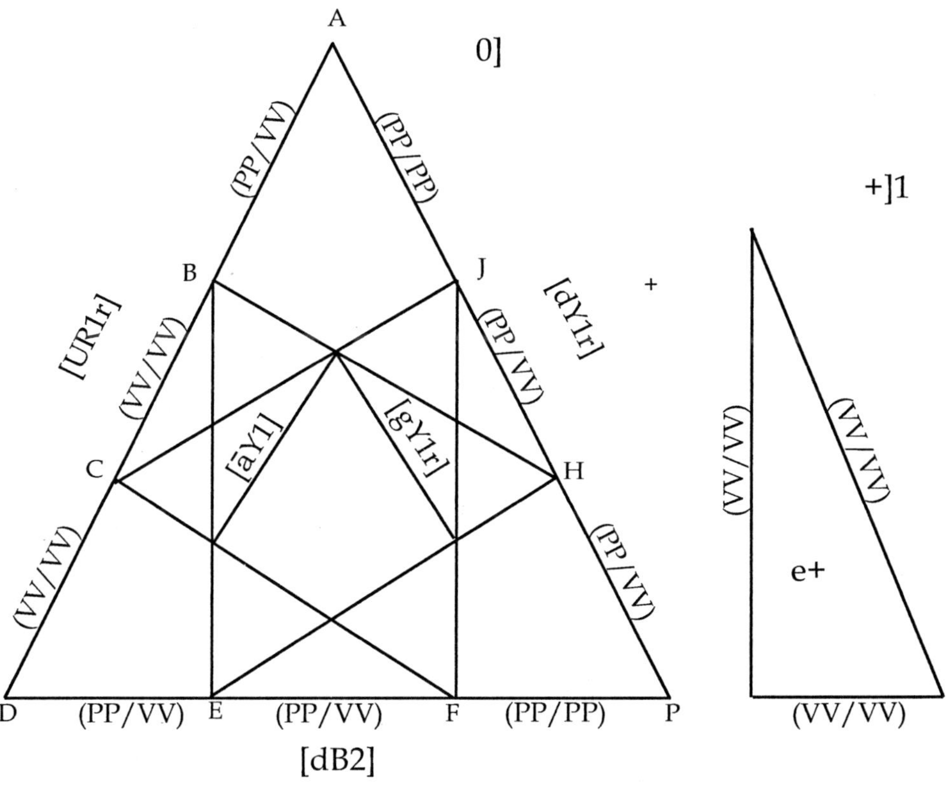

Chart XVIII.2A
(First Reaction Pathway)

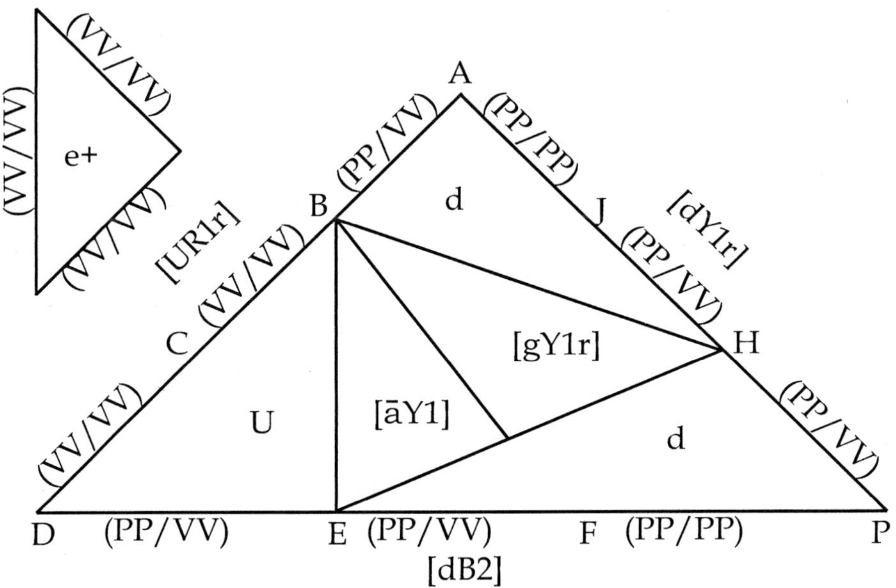

Chart XVIII.2B
(Second Reaction Pathway)

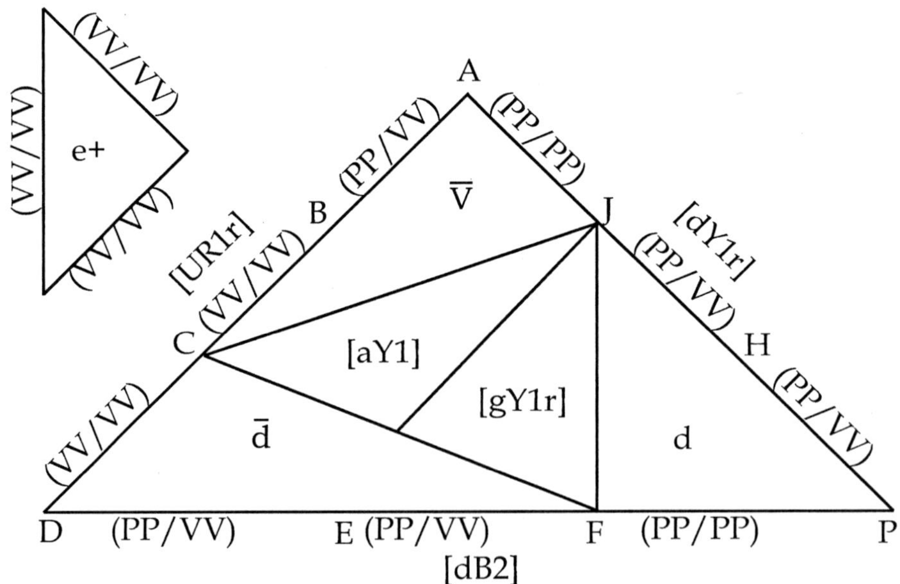

The second type of reconfiguration, due to the e+ impact, is found in Chart XVIII.2B, p.108, where a $\bar{d}$ quark, a d quark and an antineutrino ($\bar{V}_E$) form around the central g/ā meson nucleus (gY1r/āY1).

After the e+ impact on the Udd/g/ā^0 caused the reconfiguration into 2B, the e+ then further reacts with the $\bar{d}$ quark (ΔCDF) it formed, creating two up quarks, a UR1 and a UR2r, as shown in Charts XVIII.3, below. This is accomplished through the e+ impact first causing all of the boson-to-boson bonds in the $\bar{d}$ quark and the impacting e+ to be broken, thus producing six free bosons; which then reform into new boson combinations and new structures, as follows: two free W+ bosons (VV/VV) from the e+ capture a single free Z boson from the $\bar{d}$ quark (ΔCDF) to form UR1; while the remaining free W+ boson (VV/VV) in the e+ attaches itself to the remaining free W+ boson (VV/VV) and free Z boson (PP/VV) from $\bar{d}$ quark (ΔCDF) to bond together into another type of up quark, UR2r, which differs from the first up quark (UR1) in the order that its three boson constituents reformed.

These two new up quarks then attract one of the two remaining structures from Chart XVIII.2B, the d quark, dB2 (ΔFPJ),p.108.

Chart XVIII.3

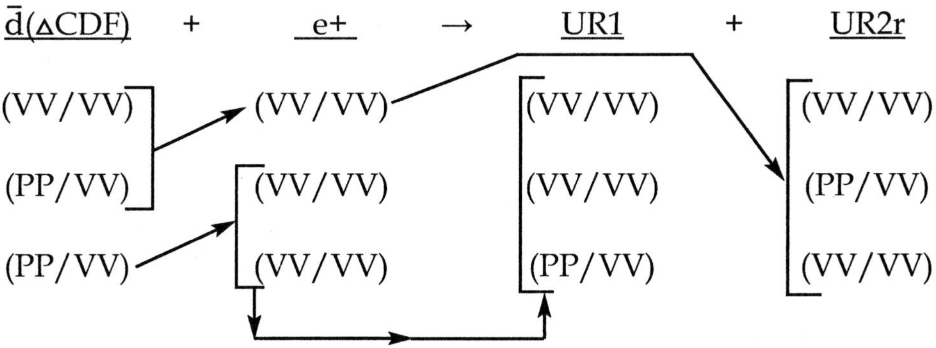

Chart XVIII.4

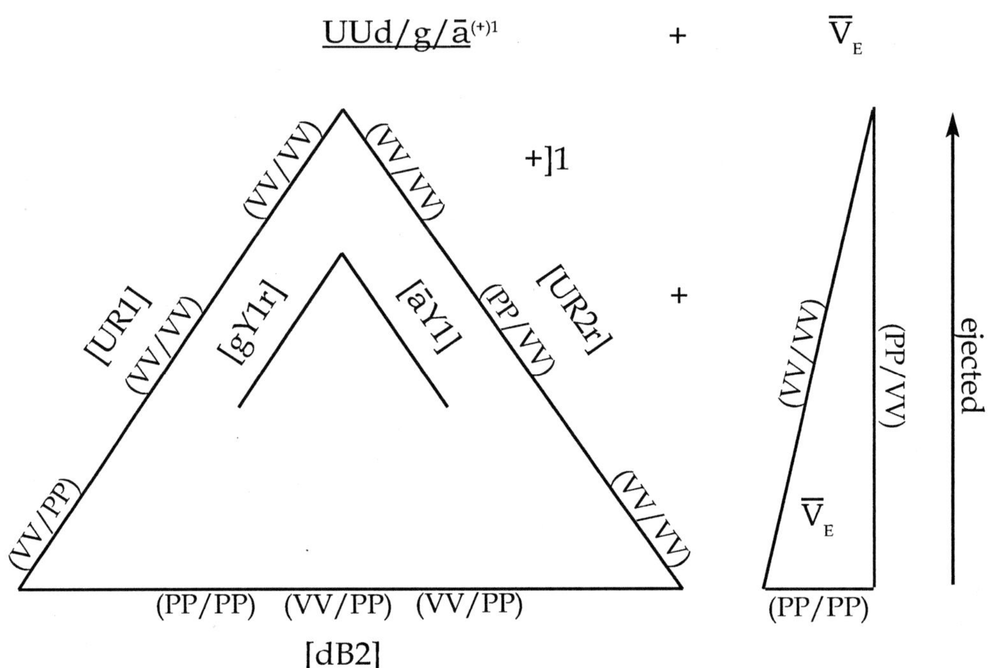

Then, these three units bond together around the unreacted g/ā meson (āY1, gY1r), producing a new UUd/g/ā$^{(+)1}$ structure; while the last remaining unit structure from Chart XVIII.2B, the $\overline{V}_E$ (antineutrino), is ejected into the surrounding space (see Chart XVIII.4), above.

Unlike the $\overline{UUd}$/g/ā$^{(-)1}$ and $\overline{Udd}$/g/ā^0 which are both eventually converted into either a proton/g/ā meson, V_E, $\overline{V}_E$, e+ or e- units by the action of the e+, all of the UUd/g/ā$^{(+)1}$ remain intact by repelling the e+, and only some of the Udd/g/ā^0 units are converted into UUd/g/ā$^{(+)1}$ and $\overline{V}_E$. The great majority of Udd/g/ā^0 units remain completely intact, saved from conversion by bonding with a UUd/g/ā^M unit to form either a deuterium $({}^{2}_{1}H)^M$ or a tritium $({}^{3}_{1}H)^M$ unit.

The latter two can then almost immediately further react with another UUd/g/ā^{M}, then add two e- units to form atomic isotopes of helium, ${}^{3}_{2}He^{0}$ and ${}^{4}_{2}He^{0}$, respectively. The latter synthesis thus saves the ${}^{3}_{1}H$ units, which are unstable and radioactive, from radioactive disintegration, by allowing their incorporation into a stable isotope of helium.

Thus, our survey of the structures based on the V photon (V) and P photon (P) has now concluded. It is as follows:

1.	+ Monopole	(VV) $^{(+)1/6}$
2.	- Monopole	(PP) $^{(-)1/6}$
3.	W+ boson	(VV/VV) $^{(+)1/3}$
4.	W- boson	(PP/PP) $^{(-)1/3}$
5.	Z boson	(PP/VV)^{0}
6.	Positron (e+)	[(VV/VV)(VV/VV)(VV/VV)]$^{(+)1}$
7.	Electron (e-)	[(PP/PP)(PP/PP)(PP/PP)]$^{(-)1}$
8.	Neutrino (V)	[(PP/VV)(PP/VV)(PP/VV)]^{0}
9.	Down quark (d)	[(PP/PP)(PP/VV)(PP/VV)]$^{(-)1/3}$
10.	Antidown quark ($\bar{d}$)	[(VV/VV)(PP/VV)(PP/VV)]$^{(+)1/3}$
11.	Up quark ($\bar{U}$)	[(VV/VV)(VV/VV)(PP/VV)]$^{(+)2/3}$
12.	Antiup quark (U)	[(PP/PP)(PP/PP)(PP/VV)]$^{(-)2/3}$
13.	Antineutrino ($\bar{V}$)	[(VV/VV)(PP/PP)(PP/VV)]^{0}
14.	Gluon/antigluon meson^{0}	[(PP/PP)(PP/PP)(VV/VV)]/[(VV/VV)(VV/VV)(PP/PP)]^{0}
15.	Proton shell/g/ā/meson $^{(+)1}$	UUd/g/ā$^{(+)1}$
16.	Antiproton shell/g/ā meson$^{(-)1}$	$\overline{UUd}$/g/ā$^{(-)1}$
17.	Antineutron shell/g/ā meson^{0}	$\overline{Udd}$/g/ā^{0}
18.	Neutron shell/g/ā meson^{0}	Udd/g/ā^{0}
19.	Proton Shell	UUd$^{(+)1}$
20.	Neutron Shell	Udd^{0}
21.	Antiproton Shell	$\overline{UUd}^{(-)1}$
22.	Antineutron Shell	$\overline{Udd}^{0}$

Of these twenty-two structures, only five will exclusively compose all of the elements of our existent universe. They are:

1. electron$^{(-)1}$ (e-);
2. proton shell/g/ā meson$^{(+)1}$, (UUd/g/ā$^{(+)1}$)
3. proton shell (UUd$^{(+)1}$)
4. neutron shell/g/ā meson^{0}, (Udd/g/ā^{0})
5. neutron shell (Udd^{0}).

We will now turn to the second half of our exposition which will once again show the evolvement of all of the preceding structures but beginning with the "Singularity" and a series of unique structures, the gravitons.

PART II

The Singularity:

The pre-existent universe, relative to the existent universe

The Singularity

Before the genesis of the present existent universe, there was only the Singularity, the absolute unified all, the "1" (One).

Then, there appeared within it a partitioned and circumscribed area, the existent universe.

Together, they were all there was or would be; a relative unified all, as well as "1".

However, to distinguish between the two epochs, we designate the Singularity after the partition event as that which is beyond the boundary of the existent universe, or beyond existence: B-e.

B-e still remains "1", since all we have done is partition it. Everything else remains the same. A single room partitioned in half is still the same single room.

B-e, therefore, also serves as the boundary, effecting a continuous and contiguous interface with the emerged, partitioned existent universe, which is itself a unified state of all that exists, or will exist, a "1".

Singularity: All ("1") and Nothing (No Thing, "0")

Prior to the partition event, the Singularity was absolutely all there was. According to the General Theory of Relativity, the existent universe appeared out of the Singularity, which was nothing (no thing) at all.

This is interpreted herein as the Singularity having a "0" spatial component; or being non-spatial as a property or attribute ("0"space).

Quantum mechanics, another relevant understanding, postulates that nothing (no thing) exists at any length shorter than

Planck length, $1.6x10^{-33}$ cm.

This is interpreted herein as defining that which is shorter, or below the Planck length, as having a "0" spatial component; or being non-spatial as a property or attribute ("0" space).

Since both of these understandings posit a "0" state, and the General Theory of Relativity explicitly defines this "0" space state as being the condition of the Singularity prior to the exact instant of the emergence of the existent universe from it, therefore it follows, in accordance with both understandings, that from the "0" spatial state of the Singularity, the existent universe emerged at Planck length in all of its three dimensions, $(1.6x10^{-33}$ cm$)^3$, in an interval of Planck time, 10^{-43} second, the smallest possible time interval designated by quantum mechanics.

Thus, both cited understandings are reconciled.

The Singularity: All and Nothing

The Singularity was absolute all and everything; also endless because it had no spatial component, thus it had no beginning or end.

Absolute all and everything because it was all and everything there was.

Absolute nothing, or no thing, because a thing, by definition, is an entity existing in time, or space. Then there was no time or space; only the Singularity.

Absolute all is rendered as "1". Even if it is infinite, the infinite "1", it is still "1".

Absolute nothing is always represented as zero ("0").

Thus, the property of being "1" and "0", simultaneously, is imprinted on all that subsequently evolves from the Singularity.

B-e, the Singularity after the epoch of the existent universe has begun, will always remain an undivided unity: a "1"; still exhibiting no spatial component, thus rendering it a "0" simultaneously. It, however, encapsulates the existent universe, while being indivisibly and inextricably bonded to it.

Thus, the existent universe is manifested or exists in the midst of nothing (no thing); or there is nothing beyond the boundary or border of our existent universe.

"1" and "0"

The Singularity was all there was: a unity or unified state, "1". Yet, at the same instant, it is nothing, "0" or "0" space.

After the partition event and its emergence within the Singularity, the existent universe appears as a myriad number of individual units , the V_E P_E gravitons ([V_E P_E]G), all of which are unified as if they were one unit; each of which, in turn, is the union of two different, single types of gravitons: the V_E graviton ([V_E]G) and the P_E graviton ([P_E]G). Nothing else exists.

These latter two gravitons, in their partnership as a V_E P_E graviton ([V_E P_E]G), appear and occupy a specific volume of space at the inception of the partition event, due to each possessing its own unique type of mass-energy matter (MEM).

The MEM of the [V_E]G gravitons is composed of its unique, non-spatial energy (V energy) inextricably and gravitationally bonded to its own unique, spatial, mass-propertied matter, or mass-matter (V mass).

The second type of MEM extant in the existent universe is that which constitutes the [P_E]G gravitons. It is constituted of its own unique, non-spatial energy (P energy) inextricably and gravitationally bonded to its own unique form of spatial, mass-propertied matter or mass-matter (P mass). No other type of MEM exists.

Thus, there are two types of energy, V and P, and two types of mass-matter, V and P.

When each [$V_E P_E$]G eventually undergoes and then completes an internal separation process, it produces the individual, separate [V_E]G graviton and [P_E]G graviton. Each of these latter two gravitons, during the separation process from each other, creates, as a by-product, untold numbers of respective photons from each of their respective substances: the V photons and the P photons.

These photons, in turn, will then self-assemble into a series of unitary structures, which will eventually constitute all of the entities found in the existent universe.

The Structure

The encapsulated existent universe, which first appears at the inception of the partition event, is also the entirety of the first of eleven gravitational dimensions, and is herein designated as G1. All eleven dimensions constitute the gravitational phase (G1 through G11).

The two essential gravitons within G1, the $[V_E]G$ and the $[P_E]G$, will evolve through these eleven gravtational dimensions, either in a merged $[V_EP_E]G$ form or as individual entities: $[V_E]G$ or $[P_E]G$.

For example, in the G7 dimension, the $[V_E]G$ and $[P_E]G$ will be completely separated from each other, thus dissolving their union as a $[V_EP_E]G$ graviton, an act of individuation for each graviton.

During the separation process, which took place in the G4 dimension, the $[V_E]G$ graviton generated V photons, while the $[P_E]G$ graviton produced P photons.

Each of the V photons has a unique magnetic charge of (+)1/12, or 1/12 of the charge exhibited by a proton, (+)1; while the P photon produces its own unique type of charge, (-)1/12, or 1/12 of the (-)1 charge projected by an electron.

The total accumulation of V photons and P photons from the foregoing process will constitute the first dimension (M1) of eleven magnetic dimensions which constitute a new phase, the magnetic phase. This M1 dimension contains all of the photons produced by all the $[V_E]G$ and $[P_E]G$ gravitons extant. These photons will evolve in and through the remaining ten dimensions of the magnetic phase into unique combinations under equally unique conditions.

Thereafter, three additional types of phases, each also containing eleven dimensions, will evolve in an ordered succession: the electric, the neutral-electric and the bond.

This five phase structure of eleven dimensions each, a total of fifty-five dimensions, is the structure of the existent universe (see Chart XX, p. 117, or the front of this book's cover).

In direct contrast to all of the different types of entities present in the fifty-five dimensions of the existent universe (gravitons, photons, protons, electrons, etc.), which are all

Chart XX

The Existent Universe: 5 Phases, 55 Dimensions

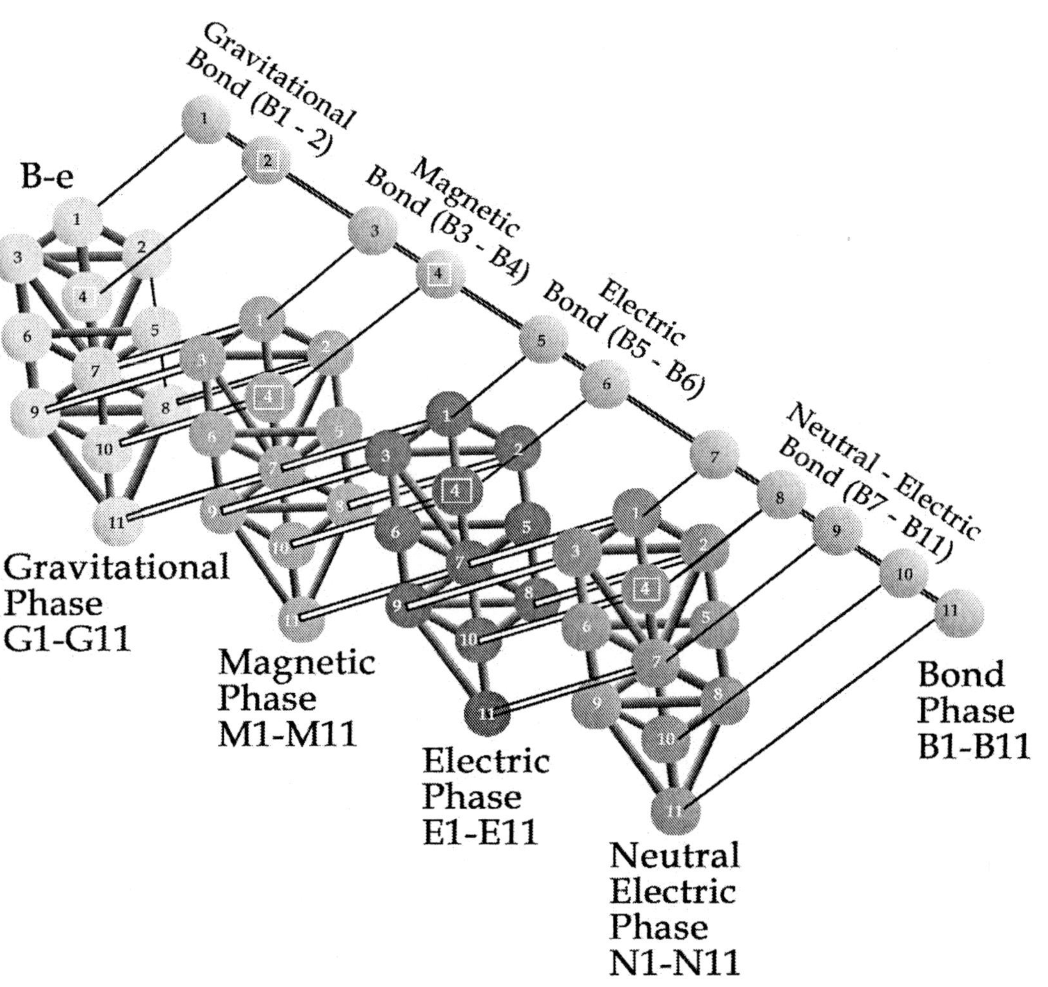

composed of spatial mass-matter inextricably bonded to non-spatial energy, the originating Singularity is totally non-spatial, "0" space, and is defined as being nothing (no thing).

B-e, the container of the existent universe, also is "0" space, as well as nothing (no thing).

Space and time are initially found in the existent universe when the gravitons become apparent, due to each suddenly acquiring a minute amount of its own unique, spatial mass-matter. Both types of gravitons are then things, having space, and each is able to be timed relative to other things.

The graviton derivatives, the V photons, and the P photons, respectively, have also been formed with the inclusion of mass-matter; thus also being things in existence.

Remerger

Remerger between B-e and the existent universe to again form the Singularity can occur any time after the partition event, but only between the absolute totality of both; essentially repeating the partition event, but in reverse.

This absolutely precludes the merger of a fractional unit of each with the other, since this would produce a "1+" or a "1-" as the resulting B-e or existent universe, respectively, destroying the unity of both as a "1".

Additionally, if the existent universe were to donate or merge a fraction of itself into B-e, that fraction would then disappear from the existent universe, tantamount to destroying a unit of mass-matter, since it would no longer be in the existent universe. This is forbidden since mass-matter can neither be created nor destroyed within the precincts of the existent universe. It can only be transformed into energy (E), in accordance with the $E=MC^2$ understanding.

Since non-spatial energy has never been isolated in the existent universe, we must therefore assume that it is inextricably bonded to mass-matter; which, in turn, and as a consequence of the $E=MC^2$ understanding, is a form of energy which we could denote as energy-mass-matter. We will, however, refer to it as mass-matter in order to distinguish it from non-spatial energy.

Non-spatial energy and mass-matter are two sides of the same coin. We will continue to explore this relationship in subsequent chapters.

PART III

Third, Fourth, and Fifth Premises

Third Premise

Our third premise states that the existent universe initially appeared out of the Singularity as myriads of spatial $V_E P_E$ gravitons ($[V_E P_E]G$); each, in turn, being the merger or union of two distinctly different gravitational substance structures, the V_E graviton ($[V_E]G$) and the P_E graviton ($[P_E]G$).

Thus, the existent universe has an equal number of $[V_E]G$ and $[P_E]G$ gravitons.

As these two graviton components of the $[V_E P_E]G$ have completely dissolved their union at the initiation of the G7 dimension, the process also caused each type of graviton to simultaneously produce from each of their unique substance bases respective V photon (V) and P photon (P) structures; which in multiple combinations will compose every single structure extant in the existent universe. The respective graviton substances are not composed of the respective photons; but photons are synthesized in and from each gravitons' own respective unique gravitational substance.

The V photons exclusively produces the (+) monopole, $(VV)^{(+)1/6}$; the W+ boson, $(VV/VV)^{(+)1/3}$; and the position, $[(VV/VV)(VV/VV)(VV/VV)]^{(+)1}$; while from the P photons are exclusively created the (-) monopole, $(PP)^{(-)1/6}$; the W- boson, $(PP/PP)^{(-)1/3}$; and the electron, $[(PP/PP)(PP/PP)(PP/PP)]^{(-)1}$.

Resulting from a combination of V photons and P photons are the following:

he Z boson	$(VV/PP)^0$
he neutrino	$[(VV/PP)(VV/PP)(VV/PP)]^0$
he antineutrino	$[(VV/VV)(PP/PP)(VV/PP)]^0$
he up quark	$[(VV/VV)(VV/VV)(VV/PP)]^{(+)2/3}$
he down quark	$[(PP/PP)(VV/PP)(VV/PP)]^{(-)1/3}$
he antiup quark	$[(PP/PP)(PP/PP)(VV/PP)]^{(-)2/3}$
he antidown quark	$[(VV/VV)(VV/PP)(VV/PP)]^{(+)1/3}$
he gluon	$[(PP/PP)(PP/PP)(VV/VV)]^{(-)1/3}$
he antigluon	$[(VV/VV)(VV/VV)(PP/PP)]^{(+)1/3}$
nd the gluon/antigluon meson	$[(PP/PP)(PP/PP)(VV/VV)/(VV/VV)(VV/VV)(PP/PP)]^0$

The more complex particles composed of V and P photons are: the neutron shell /g/ā meson^{0}, $(Udd/g/\bar{a})^{0}$; the neutron shell, $(Udd)^{0}$; the antineutron shell/g/ā meson^{0}, $(\overline{Udd}/g/\bar{a})^{0}$; the antineutron shell, $(\overline{Udd})^{0}$; the proton shell /g/ā meson$^{(+)1}$, $(UUd/g/\bar{a})^{(+)1}$; the proton shell, $(UUd)^{(+)1}$; the antiproton shell /g/ā meson$^{(-)1}$, $(\overline{UUd}\ /g/\bar{a})^{(-)1}$; the antiproton shell, $(\overline{UUd}\ )^{(-)1}$; as well as all of the elements shown in the periodic table.

All of the above complex entities are composed of some of the following: up quark $(U)^{(+)2/3}$; down quark $(d)^{(-)1/3}$; antiup quark $(\overline{U})^{(-)2/3}$; antidown quark $(\overline{d})^{(+)1/3}$; and the gluon/antigluon meson $(g/\bar{a})^{0}$; which is made up of a single gluon $(g)^{(-)1/3}$ and a single antigluon $(\bar{a})^{(+)1/3}$ in a union which exhibits a "0" charge.

Differential Existence

The newly partitioned existent universe was the near image of B-e.

The cause of its being the *near* image instead of the image is due to its having an inherent spatial property, even though it is initially of a very minute order.

The sole occupants of the existent universe at the initial point of partition are a myriad number of identical structures: the V_EP_E gravitons ($[V_EP_E]G$); all of which are bonded together into a grand unified entity: the existent universe. Nothing else exists therein.

Each of these V_EP_E graviton structures is, in turn, the union of two distinctly different and uniquely independent structures: the V_E graviton ($[V_E]G$) and the P_E graviton ($[P_E]G$). Both of these types of gravitons are each unique and an irreducible essence.

They are each a unique energy structure, not amorphous free energy, inextricably bonded to a mass-matter structure.

Each of the two gravitons is a single unified essence; a "1". Together, in a V_EP_E graviton, they are combined and operate as a single unified entity; also a "1".

The V_E graviton unit in the $[V_EP_E]G$ is caused to expand by the ignition of its inherent, cyclic force of gravity. As a result of this action, some of its own energy structure, or substance is converted into a myriad number of minute energy-matter structures, each of which we designate as an E_V-M_V complex, or V photon, which is a wave-propertied type of photon.

Another result of the $[V_E]$G expansion is the attempted compression of its attached mate in the $[V_E P_E]$G unit, the $[P_E]$G. At the conclusion of the expansion part of its force of gravity cycle, the $[V_E]$G then deflates back to its original dimensions. In the course of this latter deflationary part of its cycle and action, it pulls the $[P_E]$G into expanding its space to that volume formerly occupied by the completely expanded $[V_E]$G.

During the course of the $[V_E]$G expansion (inflation), some of the contiguous $[P_E]$G structure and substance is converted into a myriad number of new, minute, energy-matter structures, each of which we designate as an E_P-M_P complex, or P photon, which is a particle-propertied type of photon.

The primary property of the V type energy inherent in the $[V_E]$G and in the E_V-M_V complex (V photon) is self-initiated expansion caused by it's cyclic gravity force; while the P type energy inherent in the $[P_E]$G and E_P-M_P complex (P photon) causes the latter two types of structures to exhibit a property that permits them to resist and restrict the expansive and compressive action of the $[V_E]$G or E_V-M_V complex (V photon). The pull of the $[V_E]$G or the E_V-M_V complex during the deflationary part of their sine-wave function, however, cannot be resisted or restricted by the contiguous $[P_E]$G or E_P-M_P complex.

Thus, the V type energy structures are active and self-initiating; while the P type energy structures are passive, subject to activation only as a response to the actions of the V type structures. Only V type energy structures generate a cyclic force. The P type energy structures cannot generate a force; only a resistance field in response to the generation of a force.

The V type energy structure can be seen as exhibiting a nearly irresistible force, while the P type energy structure is nearly an immoveable object.

Fourth Premise

Each of the myriad numbers of $[V_E]G$ and $[P_E]G$ gravitons in the existent universe create from their own unique, inextricably bound energy-matter substance, incalculable numbers of complex structures, or complexes.

The $[V_E]G$ produce myriad numbers of $[E_V\text{-}M_V]$ complex units; while the $[P_E]G$ form myriad numbers of $[E_P\text{-}M_P]$ complex units.

The former units are V photons, while the latter units are P photons.

After each type of graviton concludes within itself the process which creates countless exact copies of its own specific type of photon complex, it then disgorges them, forming an enveloping blanket of such units which then completely surrounds each of the parent gravitons.

As the blanket of complexes grows around each graviton, it also further separates the once contiguous $[V_E]G$ and $[P_E]G$ gravitons from each other.

The act of separating the gravitons is completed in the G7 dimension of the gravitational phase, thereby causing each graviton to become individuated.

After this one time act of creation, the gravitons will never again produce these complexes (photons), since each graviton in this process has reached the minimum limit of its own necessary substance beyond which a critical instability would occur.

All of the complexes (photons) thus created, however, will continue to expand in size over the course of the ensuing billions of years, eventually attaining size levels countless billions of times their original generated size level and that of their progenitor gravitons.

The actual mechanism of expansion will be detailed as each of the two types of complexes is described in the following sections:

I. The $[E_V\text{-}M_V]$ type of complex is created solely in the $[V_E]G$, as single, unified entities, which are also called V photons.

At the conclusion of a process within the $[V_E]G$ part of the $[V_EP_E]G$ graviton, which can be considered as a type of gestation period, all of the [E_V-M_V] units produced therein are disgorged, thereby surrounding the parent graviton, while also coming into an equilibrium condition.

Each [E_V-M_V] photon unit contains untold numbers of E_V energy units, which are stored within the complex. Each E_V unit is always inextricably and magnetically bonded to all of the M_V type units which are also present and stored within the photon complex.

Since the E_V units are "0" space, an almost infinite number can be kept within each individual [E_V-M_V] photon unit. When a single non-spatial E_V unit is eventually transformed into a spatial M_V type unit, the [E_V-M_V] photon unit which houses it expands by the exact same spatial amount as the E_V unit expands by in transforming itself into its new form, or state, the M_V unit.

In order for this expansion to take place, an E_V unit must transform its non-spatial energy into spatial mass-matter, in accordance with the $E=MC^2$ understanding.

In a first step toward achieving this end, the non-spatial E_V unit transforms and appears as a mass-matter unit, which we now label as an M_V^F type of unit, in the shortest interval of time, 10^{-43} second (Planck time). This is due to the ignition of it's cyclic magnetic force. At this initial instant of possible visible detection, its length in each of its three dimensions is $1.6x10^{-33}$ cm, the Planck length (PL), since any shorter distance is nothing, or "0" space, according to quantum mechanics. Its size, therefore, is $(PL)^3$, or $(1.6x10^{-33}\text{ cm})^3$.

After this first instant of appearance and detection, a second step is then initiated by it's magnetic force that further expands or multiplies the $(PL)^3$ sized unit by an expansion factor of 9×10^{20}cm**. In equation form these steps are rendered as:

$$(1)\quad E_V \xrightarrow{\text{appears as}} M_V^F\ (\text{at }(PL)^3{*})$$
$$(2)\quad M_V^F\ (\text{at }(PL)^3{*}) \xrightarrow{\text{expands to}} M_V^F\ (\text{at }(PL)^3{*} \times (9 \times 10^{20}\text{cm.})\ {**}$$

Once the above unit at $(PL)^3$ size has expanded to its new $(PL)^3xC^2$ size, a third step is then immediately initiated as the M_V^F unit is caused to reverse its action due to it's sine wave function, and deflate back to its original $(PL)^3$ size, as demonstrated in the following:

* $(1.6 \times 10^{-33}\text{cm.})^3$

** 9×10^{20}cm. is the number part of C^2; used herein as the expansion factor (multiplier).

(3) M_V^F (at $(PL)^3 x (9 x 10^{20}$cm.)) $\xrightarrow{\text{deflates to}}$ M_V^S (at $(PL)^3$)

The deflation is a result of the second, deflationary part of the sine wave function the magnetic force follows: first, from "0" value, reaching its maximum force intensity; then secondarily declining back to its former value of "0".

In (3) above, we rename the M_V^F in its final deflated form as an M_V^S, due to a change in its size and property.

The nomenclature used in distinguishing the mass-matter expanded M_V^F unit from the deflated mass-matter M_V^S unit is based on the "flexibility" property inherent in the M_V^F unit; thus the "F", denoting this property, above the "V", which indicates the unit's V type mass-matter base.

The M_V^S unit, on the other hand, once formed, can no longer expand, or deflate further to return to its original "0" space E_V status. It is "spent", and therefore we append an "S" over the "V" to denote this property.

This spent M_V^S unit it is then stored in the $[E_V\text{-}M_V]$ photon complex unit (V photon).

Each time a non-spatial E_V unit is transformed into its final form as a spatial and spent unit, the M_V^S unit, it has expanded by $(PL)^3$, or $(1.6x10^{-33}\text{ cm})^3$, as does the $[E_V\text{-}M_V]$ complex (V photon) containing it, as well as the entire existent universe.

The above E_V transformation is based on an ever-repeating magnetic force cycle present within all E_V units. Each time the cycle and its magnetic force is generated, a single E_V unit is triggered into completing the conversion process into an M_V^S unit in each $[E_V\text{-}M_V]$ complex or V photon unit extant in the existent universe.

Therefore, if one multiplies $(PL)^3$ by the number of E_V units present in each $[E_V\text{-}M_V]$ photon unit extant at the moment of disgorgement from the $[V_E]$ graviton, and then multiplies this result by the number of $[E_V\text{-}M_V]$ photon units then present in the existent universe, we arrive at the total expansion value potential possible for this E_V factor; one of three factors which contribute to the maximum possible expansion size of the existent universe.

Although a $(PL)^3$ spatial volume increase in a single $[E_V\text{-}M_V]$

unit is quite minute, in terms of all such units present in each $[E_V\text{-}M_V]$ photon and all $[E_V\text{-}M_V]$ photon units extant in the universe, it is enormous.

In terms of the very large numbers of E_V present in each $[E_V\text{-}M_V]$ complex or V photon, untold billions in number, the expansion of all of these E_V units in a single V photon would expand its size by untold billions over the many billions of years the cyclic expansion process would take for its completion.

In order, however, to determine the potential total magnitude of expansion possible for the entire existent universe from the point of its initiation, we would have to add two additional expansion magnitudes to the preceding number:

(1) The volume of space attained by all the $[V_E]$G and $[P_E]$G gravitons after disgorgement of all of their complexes; and

(2) The volume of space created by all of the $[E_P\text{-}M_P]$ complexes (P photons) extant due to the conversion of all of the E_P units present within each into M_P units.

In order to determine the latter, we will turn to an exposition on the $[E_P\text{-}M_P]$ complex, as follows:

II. The $[E_P\text{-}M_P]$ type of photon complex is only produced in the $[P_E]$G, as a single, unified entity; it is also referred to as P photon.

The process of creating the $[E_P\text{-}M_P]$ photon complex begins with a force of gravity wave being generated in the $[V_E]$G graviton, which is attached to or merged with a $[P_E]$G graviton, as a (V_EP_E) graviton.

The force of gravity is due to a difference in stability between the absolute stability present in the Singularity, out of which the $[V_E]$G appeared at the partition event as part of the V_EP_E graviton, and the relative instability of the substance of the $[V_E]$G in the $[V_EP_E]$G, due to the presence of mass-matter in it after partition.

The Singularity is devoid of mass-matter, while the $[V_E]$G has initially a very minute amount of a mass-matter component, inextricably bonded to its non-spatial energy component.

This small content of mass-matter is what causes a gravity force wave to be generated in each $[V_E]G$; which then expands its substance, resulting in the creation of $[E_V-M_V]$ complexes which move to its exterior surface area. These complexes also concentrate a very high percentage of the mass-matter contained in the $[V_E]G$.

Thus, when the $[E_V-M_V]$ photon units are finally disgorged from the $[V_E]G$, they form an enveloping layer around and outside the $[V_E]G$, which also then leaves the $[V_E]G$ with a substantially lower concentration of mass-matter, and thus a concomitantly much greater stability in its remaining substance. In this way, the $[V_E]G$ increases its stability and thus, precludes generating its gravity force cycle.

However, the $[V_E]G$ pays a price for this greater stability, in that in order to produce these $[E_V-M_V]$ units, it has reached the very lowest limit of energy and mass-matter substance permitted in the confines of it's body, beyond which catastrophic instability would ensue. Thus, it will never again produce $[E_V-M_V]$ units.

In the actual $[E_V-M_V]$ creation process, a cyclic gravity force wave is generated within the $[V_E]G$, increasing in strength from a "0" point to its maximum peak, causing a concomitant maximum inflationary expansion of the substance of the graviton.

During the course of the above cyclic gravity force manifestation process in the $[V_E]G$, some of its substance is transformed into $[E_V-M_V]$ complexes.

During the above actions, the $[P_E]G$ unit, contiguous to the $[V_E]G$ in a $[V_EP_E]G$ graviton, is impinged upon by the compressive force generated by the $[V_E]G$'s expanding outer surface area, which is now exclusively composed of newly created $[E_V-M_V]$ units.

The $[P_E]G$'s substance responds to this expansive thrust by the $[V_E]G$ by generating $[E_P-M_P]$ complex units at its own outer surface area, where they exhibit an extraordinary property of rigidness. Thus, the $[P_E]G$ is able to completely resist the impinging force of the $[V_E]G$ by being able to generate its own rigidity response units to offset the force.

This reaction to the $[V_E]G$ expansion on the part of the $[P_E]G$ is totally passive. As the $[V_E]G$'s active gravity force manifestation increases, so does the number of $[E_P-M_P]$ complex units generated in the $[P_E]G$; and thus a concomitant increase in rigidity or stiffness in the $[P_E]G$, thereby canceling out any intrusion by the $[V_E]G$ into the substance of the $[P_E]G$.

As the expansion of the $[V_E]G$ reaches its maximum, with a concomitant high concentration of $[E_V\text{-}M_V]$ units at its outer surface area, the contiguous $[P_E]G$ also reaches its maximum number of $[E_P\text{-}M_P]$ units at its outer surface area, as well as a maximum rigidity in opposition to the $[V_E]G$ expansive force.

At that point, the two opposing contiguous surface areas suddenly shatter into distinct and free $[E_V\text{-}M_V]$ and $[E_P\text{-}M_P]$ units, respectively.

At a Planck time interval thereafter (10^{-43} second), the gravity force begins to decline, accompanied by a concomitant deflation of the spatial volume of the $[V_E]G$ unit.

This causes the new $[E_V\text{-}M_V]$ and $[E_P\text{-}M_P]$ units to flow out of their respective gravitons, producing an interface area between the two types of gravitons composed of these units, which then move to entirely surround their respective gravitons.

During the ensuing period of gravity force decline and deflation in the $[V_E]G$ spatial volume to an eventual "0" gravity force value and a $(PL)^3$ spatial volume, the $[V_E]G$ pulls (or attracts) its contiguous $[P_E]G$ mate into expanding its substance, thus inflating the $[P_E]G$ volume of space to exactly the volume of space previously occupied by the $[V_E]G$ at its maximum point of expansion.

Underlying this process is the passive nature or property of the $[P_E]G$ and all structures that eventually derive from it (W- boson, electron, etc.). These types of structures cannot generate a force, be it gravity, magnetic or electric, as can the $[V_E]G$ and its derivative structures (W+ boson, positron, etc.). The $[V_E]G$ is categorized as active; the $[P_E]G$ as passive. This categorization also applies to their respective derivatives.

Thus, the passive $[P_E]G$ can only be pulled by the $[V_E]G$ into expansion, thereby concomitantly filling the actual space the $[V_E]G$ is vacating as it deflates.

Such action also prevents a hole in space from developing due to the deflation of the $[V_E]G$ substance, since such a hole is forbidden in that its contents would be a new type of matter, unrelated to either

$[V_E]G$ or $[P_E]G$. In the existent universe, matter can neither be created nor destroyed, only transformed from one form into another as understood by the $E=MC^2$ equation.

The Understanding

All of the foregoing described actions and reactions by the gravitons or their creations, the complex type of units (photons), are entirely governed by the $E=MC^2$ understanding.

For example, when a V type energy, "0" space, E_V unit in an $[E_V\text{-}M_V]$ complex (V photon) generates an internal magnetic force wave, which is governed by a sine wave function, it first transforms into a V type, mass-matter unit, M_V^F, which then appears and exhibits a spatial volume of $(PL)^3$ (see A. below).

As a now mass-matter M_V^F unit, it still, however, must perpetuate the remainder of the magnetic force sine wave, thereby having to further expand and thus multiply its $(PL)^3$ space by an expansion factor of (9×10^{20}cm.) (see B. below). All of the oregoing actions are embodied in the following variant of the $E=MC^2$ understanding:

A. $E_V \xrightarrow{\text{appears as}} M_V^F$ (at $(PL)^3$)

B. $M_V^F \xrightarrow{\text{expands to}} M_V^F$ (at $(PL)^3 x(9 \times 10^{20}\text{cm.})$)

Here, in A., above, the energy-based, non-spatial E_V unit has been converted by the generation of its own magnetic force wave (→) into a mass-matter, spatial unit, initially occupying a $(PL)^3$ volume of space, the M_V^F .

As an M_V^F , it then further generates and perpetuates it's magnetic force wave action, which concurrently causes an expansion factor, (9×10^{20}cm.), to come into play, causing the M_V^F to then inflate to a maximum expansion value of $9 \times 10^{20}\text{cm} \times (PL)^3$.
This seen in B., above.

When the M_V^F unit in the $[E_V\text{-}M_V]$ complex (V photon) finally reaches this $(PL)^3x(9 x 10^{20}cm.)$ maximum, it then immediately begins to deflate back to its initial $(PL)^3$ spatial volume, due to having to follow its magnetic force sine wave function. This causes an E_P, non-spatial, P type energy unit contained in a contiguous $[E_P-M_P]$ complex (P photon) to be pulled, or attracted, by the declining force of magnetism in the $[E_V-M_V]$ unit into expanding; thereby transforming the E_P into a new form M_P^F, a P type mass-matter, spatial unit; the "F" above the bottom "P" indicating it is a flexible unit capable of being pulled, and thus expanded; while the bottom "P" designates the unit as being derived from the $[P_E]$G and being composed entirely of P type energy that has been transformed into P type mass-matter.

Once the E_P to M_P^F transformation step is completed, in response to the initial deflation of the M_V^F unit in the V photon, the new M_P^F then has a spatial volume of $(PL)^3$. It is then pulled into further expansion by the continuous deflation of the M_V^F unit in the $[E_V-M_V]$ complex, as it follows its sine wave function to zero value. The M_P^F, in the $[E_P-M_P]$ complex (P photon), will eventually reach a maximum value of $(PL)^3 x(9 x 10^{20}cm.)$, the same inflationary spatial volume that had been attained by the M_V^F in the $[E_V-M_V]$ complex unit just prior to beginning the deflationary part of its sine wave function.

At this point of maximum space volume attainment, the M_P^F unit within the $[E_P-M_P]$ complex (P photon) is also at its maximum in rigidity, and will continue to remain in this condition, as well as holding its $(PL)^3x(9 x 10^{20}cm.)$ space volume, so long as there is one unreacted E_V unit in a contiguous $[E_V-M_V]$ complex which is then able to initiate and generate its own magnetic force sequence. Only one E_V unit, within a V photon, at any one time, is permitted to generate a magnetic force wave. At the end of its sine wave cycle, as a transformed M_V^S unit, another E_V unit will instantly generate a magnetic force. Thus, the magnetic force will always be present and continuous as long as there is one E_V unit present in a V photon to generate it.

The M_P^F unit in its final rigid form will now be designated as an M_P^S unit, indicating its new property of rigid "solidity", in contradistinction to its previous form as a M_P^F, where it was quite flexible and elastic in property in order to be capable of being passively expanded by$(9 x 10^{20}cm.)$; or being able to be pulled into expanding its substance by a factor of (9 x 10^{20}cm.). As a M_P^S unit, it is spent, unable to deflate as long as an E_V unit is present in a contiguous [E_V-M_V] complex and is generating a magnetic force.

Thus, the total value of the expansion of a single E_V in an [E_V-M_V] complex (V photon) and a single E_P in an [E_P-M_P] complex (P photon) is, therefore, $(PL)^3$+[$(PL)^3$x(9 x 10^{20}cm.)].

Since there are equal numbers of V and P photons in the universe, one can then multiply the total number of pairs of V and P photons extant by the (($(PL)^3$+$(PL)^3$x(9 x 10^{20}cm.)) value in order to arrive at the total amount of expansion of spatial volume the universe can achieve during the conversion of one E_V unit and one E_P unit in each of their photons into mass-matter M_V^S and M_P^S units, respectively.

Since there are initially equal numbers of non-spatial E_V and E_P units in each respective V and P photon extant, then if one multiplies the foregoing total amount of expansion achieved in the above paragraph, by the total number of either E_V or E_P units contained in each of their respective photons at the initial moment of equilibrium after their disgorgement from their respective gravitons, one arrives at the total expansive volume of space possible for the universe to achieve due to its V and P photon content. Thus, E_V and E_P are "dark energy"; and M_V^S and M_P^S are the "dark matter" the E_V and the E_P will be converted into.

Additionally, the number of graviton pairs extant in the universe has had only one expansion episode, therefore the value of the expansive volume of space for each pair is [$(PL)^3$ + $(PL)^3$ x (9 x 10^{20}cm.)].

Multiplying this value by the number of graviton pairs extant results in their achieved expansive volume of space value which, when added to the expansive value just calculated above due to its photon content, gives the finite size volume of space to which the universe can maximally expand.

Although the calculation of this value may be useful, the fact that it signifies that the universe is finite is more important. The existent universe, then, is closed. It is enclosed by B-e, which has no spatial volume. Therefore, the existent universe exists in the center of nothing (no thing), "0" space.

END GAME

An intriguing question arises as a consequence of the preceding exposition: What will happen when all of the non-spatial E_V and E_P energy units in each of the respective V photons and P photons extant are completely converted into their respective spatial, mass-matter equivalents, the M_V^S and M_P^S units?

Since the V and P photons exclusively compose all of the elements, stars, planets and space dust, what then becomes of the entire existent universe at a point in time when there is a sudden and complete absence of E_V and E_P units?

The answer purposes the eventual outcome of the existent universe.

Only each of the myriad number of E_V units stored in each V photon has the capability of generating an expansive magnetic force. This force is propagated throughout the process previously described in which the E_V non-spatial unit is converted into a spatial M_V^S unit within the V photon.

As a result of this process, a parallel transformation process takes place in the contiguous P photon, in which E_P, non-spatial units, are converted into spatial M_P^S units.

The above two cyclic processes have been in effect for the past fifteen billion years and will continue until the supply of E_V and E_P units runs out. They are and will be mainly responsible for the expansion and size of the existent universe until the exhaustion point of all E_V units in all V photons is reached.

This continuous cyclic magnetic force generation has resulted in all M_V^S units in the V photons and all M_P^S units in the P photons, since their disgorgement from their respective gravitons, to each remain at a constant spatial volume of $(1.6 \times 10^{-33}$ cm$)^3$ and $(1.6 \times 10^{-33}$ cm$)^3 \times (9 \times 10^{20}$cm.), respectively.

In essence, the continuous cyclic magnetic force generation causes the ever-expanding volume of space, as well as not permitting a collapse of both photon types, as well as the Universe.

When, however, at the instant the last E_V unit in each V photon extant has been converted into an M_V^S unit, with a concomitant conversion of the last E_P unit in each P photon extant into an M_P^S unit, the magnetic force ceases to exist, since no E_V unit is extant to generate it.

At that instant, there is also an immediate and complete loss of the passive rigidity response in all M_P^S units in all P photons extant, since it is only the magnetic force which causes this passive resistance response in the form of rigidity in the first place.

This complete loss of rigidity then results in the sudden collapse and deflation of all the P photons' M_P^S units extant from a size value of $(1.6x10^{-33}\ cm)^3 \times (9 \times 10^{20} cm.)$ to their original appearance size value of $(1.6x10^{-33}\ cm)^3$.

The velocity generated by this sudden collapse, however, is so great, due to the complete absence of any arresting magnetic force, that the value of the volume of space occupied by the M_P^S unit actually drops to below $(1.6x10^{-33}\ cm)^3$, the minimum value for a spatial mass-matter unit (M_P^S) to be mass-matter and existent.

Thus, all of the M_P^S units are now E_P, non-spatial energy units, due to their being below the minimal $(1.6x10^{-33}\ cm)^3$ spatial volume.

This sudden fall in spatial volume of all the M_P^S units to zero space, and thus the total spatial volume of all the P photon units to zero as well, has also caused a concomitant, powerful pull on all of the M_V^S units in all of the V photons extant, resulting in a deflation in their spatial volume to also just below $(1.6x10^{-33}\ cm)^3$, or zero space, which results in their instant conversion into non-spatial E_V units again.

Thus, the entire existent universe at this instant has devolved completely into E_V and E_P non-spatial energy units, in their respective V and P photon units. Therefore, all of the V and P photons are composed exclusively of respective E_V and E_P non-spatial energy, and are "nothing", or zero space. The existent spatial, photonic universe will now have disappeared, and only the gravitons are still apparent and present as spatial units.

Then, each individual type of graviton will reabsorb the exact same number of respective non-spatial V and P photons as each had generated as spatial V and P photons during disgorgement at the initiation of photonic universe (G7).

Thereafter, each graviton then repeats the exact same sequence of steps and processes they originally executed from the G1 to G7 dimensions of the gravitational phase, but in reverse, including the penultimate step of merging all of the $V_E P_E$ gravitons together. Then, the final step is executed where the unified $V_E P_E$ gravitons merge with B-e, reversing the partition event, resulting once more in the Singularity.

A "0" Space, Energy Entity

For most, the very concept of a "0" space or non-spatial entity is seemingly impossible to accept, especially since our everyday experiences in our world appear to be exclusively dominated by spatial entities.

Even if such a "0" space entity were present, its detection would be entirely dependent on the exclusively spatial, mass-matter sensing organs of our five senses.

Having to rely only on these spatial five senses, however, would preclude the discovery and identification of a wholly "0" space entity.

It is only our sixth sense, the mathematical, which permits the visualization and comprehension of such "0" space entities, together with the realization that they are transformable into space occupying forms and vice versa.

This understanding has been permanently engraved into our tradition of science as $E=MC^2$.

Here, in this icon, two independent and radically different forms, "E" and "M", non-spatial ("0" space) and spatial, respectively, are visualized mathematically as two different manifestations of the same essence; while the equal sign (=) acts as the road between their two distinct forms.

Yet, although the "E" is apparent and displayed in the icon, not one drop, or even a (Planck length)^3 vessel of it, has ever been isolated for even an instant of Planck time (10^{-43} second) in the entire history of the existent universe.

However, there it is: "E", in probably the most important icon in science and its existence sworn to by its most honored practitioner.

In what is the most honored understanding in all of the

sciences, the General Relativity Theory, Einstein unequivocally states that the existent universe emerged out of a Singularity which was nothing!

No one, to date, has effectively contradicted this.

However, is the above "nothing" therefore "something"?

No. Nothing is not a thing.

It, however, is.

The understanding, herein, is based on energy, E, not having a spatial component while in this E form. It is "0" space.

Yet, we know that it can and does transform and manifest into a spatial, mass-matter form, M.

Thus, being able to manifest M from E is obviously a latent presence in the E form, which cannot be detected by any of the mass-matter senses.

Furthermore, energy is obviously existent, but apparently is inextricably bound to spatial mass-matter which it cannot be separated or isolated from as a distinct and independent entity.

Our only knowledge of the actual and definitive existence of energy is in its demonstrated actions and reactions found in such manifestations as muscles twitching, hearts palpitating and bombs exploding. In fact, there is not one instant in our lifetime that the energy manifestation is not present.

In this presentation, notable non-spatial "0" space energy entities such as E_V, E_P and the Singularity have been purposed, as well as their inextricably bound combinations with mass-matter: [E_V-M_V], [E_P-M_P] and [B-e/Existent Universe].
Each is. All are.

Is Nothing Something?

We shall herein purpose two possible understandings of what a "0" space entity, such as an E_V or E_P, is.

Both are based on quantum mechanics and its postulate that nothing exists below the Planck length of $1.6\text{x}10^{-33}$ cm.

This would then define the smallest possible three-dimensional spatial volume to be $(1.6\text{x}10^{-33}\text{ cm})^3$.

Thus, an entity may be nothing, or "0" space, if each of its three-dimensional lengths is less than $1.6x10^{-33}$ cm and are all approaching "0" length, and yet can never reach it.

In this first understanding, a state exists where each of the three-dimensional lengths may be $1x10^{-43}$ cm, or $1x10^{-53}$ cm, or $1x10^{-153}$ cm, or lower, with no absolute minimum size limit.

Such a state, with an entity having all three of its dimensional lengths below $1.6x10^{-33}$ cm, but never reaching "0" length, can generically be referred to as nothing or "0" space. An entity, an E_V or E_P, in this state is a "0" space, or non-spatial entity.

In a second possible understanding, an E_V or an E_P entity can also be defined as "0" space, or non-spatial entity, if each of its three-dimensional lengths goes below $1.6x10^{-33}$ cm, instantly reaching a length that is absolute "0".

Thus, such an entity is also "0" space; but, more correctly, an absolute "0" space entity.

To most, the latter understanding stretches the intellect to the absolute limit of its imaginative capacity, except for that which can be expressed and realized mathematically.

In the former understanding, however, there is always something present in the "0" space entity, even if it consists of only an infinitesimal wisp of something spatial. Total emptiness, here, does not exist.

The choice of which of the above two concepts is correct, however, remains with each reader.

None of the foregoing, however, relates in any way to the Singularity (B-e), since nothing like the Singularity has, is, or ever will be in the existent universe.

Thus, no description of the Singularity is possible, except that its latent attributes are manifest in the existent universe. But even such attributes, in toto, cannot describe the Singularity. One cannot describe in any manner, shape or form what has never been in the existent universe.

All that can be said is: The Singularity is the Singularity; there is nothing (no thing) like the Singularity in existence.

Fifth Premise

The present existent universe has evolved over a period of approximately fifteen billion years, through five successive, separate and overlapping phases; each phase, in turn, progressing through a successive series of eleven dimensions, for a total of fifty-five dimensions, each dimension exhibiting a unique set of parameters.

The five phases, in order of evolvement, are as follows:

	Phase	Abbreviation
1.	Gravitational	G
2.	Magnetic	M
3.	Electric	E
4.	Neutral-Electric	N
5.	Bond	B

(See Chart XX, p. 117.)

The first four phases are generated in successive order: 1. gravitational (G), 2. magnetic (M), 3. electric (E) and 4. neutral-electric (N).

The fifth phase, bond (B), however, is created in part by each of the first four phases.

The gravitational phase's G1 and G4 dimensions generate the bond phase's B1 and B2 dimensions, respectively; while the magnetic phase's M1 and M4 dimensions create the respective B3 and B4 dimensions. The third phase, the electric, and its E1 and E4 dimensions produce successively: the B5 and B6 dimensions, respectively. While the fourth phase, the neutral-electric, adds from its N1, N4, N7 N10 and N11 dimensions: B7, B8, B9, B10 and B11, respectively, a total of five additional bond dimensions; a grand total of eleven bond dimensions.

All of the above eleven bond generating dimensions have bonds present between various entities within their precincts. These bonds lock and weave these entities together into a network which permits unified responses across all dimensional lines, as well as keeping the universe and its myriad entities and systems in a set equilibrium.

Entity and Bond Dimensions

Each of the eleven bond dimensions, illustrated in Chart XX, p.117 (B1-B11), is generated by one of the following eleven respective entity dimensions (G1, G4, M1, M4, E1, E4, N1, N4, N7, N10, N11); each of which is in the central column of Chart XX.

An entity in each of these eleven entity dimensions , is, however, affected by and responds to changes in a bond in any one of the eleven bond dimensions.

Each of the eleven bond dimensions is interlinked with all of the others, producing a type of network which requires that the action of a single entity in any one of the eleven entity dimensions be balanced by the reaction of another entity in the same dimension or in one of the other entity dimensions.

Thus, for example, the spin of one electron must somewhere be balanced by the spin of another electron, even if both are separated by a great distance, or are in different dimensions. Once effectuated, the equilibrium condition in the bond dimensions returns. This is due to all fifty-five dimensions of the existent universe being a free-flowing continuum and not a group of partitioned and rigidly demarcated dimensional areas having distinct barriers between them.

This can be better understood in light of the fact that the first entity dimension, G1, initially contained only a single entity, a unified structure composed of all of the $[V_E P_E]G$ gravitons extant, bonded together by a single gravitational bond in the B1 dimension.

All of the entities in the remaining ten entity dimensions which generate a bond dimension in Chart XX are distinct parts of this original unified particle which inaugurated the existent universe in G1.

Therefore, it follows that all the bonds in these ten respective bond dimensions are each a fraction of the initial gravitational bond generated in the B1 bond dimension.

This continuum of eleven bond dimensions, therefore, can also act to signal an alteration in a entity's bond in one bond dimension to another entity's bond in the same, or another, bond dimension in order to initiate a mitigation of the consequences of that bond alteration by the use of an available mechanism.

Thus, a governing principle, arising out of this continuum of entities and interlinked network of bond dimensions, is that every action is accounted for by an equal and opposite reaction. This results in the maintenance of an absolute unified universe in which an inherent absolute equilibrium is always maintained.

This single, unified entity, composed of all the myriad numbers of V_EP_E gravitons ($[V_EP_E]$G) extant in the now partitioned existent universe, the G1 dimension, then immediately begins to have its B1 gravitational bond decline in strength, eventually leaving each of the myriad number of $[V_EP_E]$G gravitons free and unbonded at the conclusion of processes in the G4 dimension.

G4 is a process dimension, which is unlike all the other ten dimensions composing the gravitational phase. It has at least two processes proceeding simultaneously in it's precinct, as well as all of it's entities being fugitive and thus not capable of being isolated.

In one of the G4 processes, the gravitational bond holding together the $[V_E]$G and $[P_E]$G graviton members of the individual, free $[V_EP_E]$G graviton eventually declines. This causes the $[V_E]$G and $[P_E]$G to be able to separate from one another and become individuated entities.

In a second concurrent process, the $[V_E]$G and the $[P_E]$G each initiate an internal transformation in which each one transmutes some of its substance into a new type of entity, the V photon and the P photon, respectively.

Each of these V and P photons also has transferred to it the bond-forming capacity of their $[V_E]$G and $[P_E]$G progenitor's substance, in relative proportion to the fractional amount of its share of the total gravitational substance.

The bond-forming capacity or charge is as follows: 1) the V photon has a (+) 1/12 charge; 2) the P photon carries a (-) 1/12 charge. The former charge is 1/12 of the (+) 1 charge exhibited by the positron or proton; while the latter is 1/12 of the charge exhibited by an electron, (-) 1.

Each of these two types of photons were created from the substance of their progenitor graviton entities and are now differentiated from each of their progenitor entities by being referenced as magnetic entities, instead of minute gravitational entities; as well as being transferred to and occupying a new area, the M1 dimension of a new phase, the magnetic.

Yet, the V and P photons are still minute gravitational entities exhibiting respective gravitational force, field and bond properties; which, because of their relative minute nature have been inaccurately deemed to be magnetic entities, with magnetic forces, fields and bonds, as if they were seemingly unrelated to their true gravitational progenitors. **Magnetic is simply minute gravitational.**

As electric particles are composed exclusively of magnetic entities, **electric therefore also means minute gravitational** in respect to it's type of entities, as well as to the true nature of the electric force, field and bond.

Thus, all entities, forces, fields and bonds extant in the existent universe are gravitational; and thus a grand unified state, or a state of grand unification, exists in our universe.

We will,however, retain the historical momenclature of gravitational,magnetic and electric throughout this text for continuity in our presentation.

Magnetic Bonds

The rule governing bonding in all eleven dimensions of the new magnetic phase is: like particles attract, opposites repel; the reverse of the rule in electric dimensions.

This rule causes $V^{(+)1/12}$ photons and $P^{(-)1/12}$ photons to repel each other; then flow into their own respective and totally segregated areas of the M1 dimension.

Following the bonding rule, as previously outlined for magnetic dimensions, pairs of $V^{(+)1/12}$ photons bond with each other to create $(VV)^{(+)1/6}$, (+) monopoles; while pairs of $P^{(-)1/12}$ photons bond, producing $(PP)^{(-)1/6}$, (-) monopoles.

Thus, no $(VP)^0$ type monopole structure can be produced due to the opposites repel part of the bonding rule in magnetic dimensions. The closest we get to such a structure is the Z boson, $(VV/PP)^0$; which can only be created in the electric phase, where opposites attract, and therefore can bond.

Each of the two types of magnetic monopoles then flow into their own respective and segregated areas. Together, these two new areas now constitute a new magnetic dimension, the M4, which like the G4 is a process dimension. These two areas (G4 and M4), along with the E4, N4 and B2, are unique; as they are the only process dimensions of the total of fifty-five dimensions constituting the entirety of the existent universe. The first four contain respective structures too fugitive to be isolated: P_E V_E gravitons in the process of seperating; monopoles; gluons; antigluons; quarks and double-quarks; while the fifth, the B2, contains the gravitational bond still splicing together the $[P_E]G$ and $[V_E]G$ in the P_E V_E graviton as they are in the process of seperation.

M4 Reactions

In the (+) monopole sector of the M4, pairs of $(VV)^{(+)1/6}$ attract and bond, producing a new type of unit, the electric W+ boson, $(VV/VV)^{(+)1/3}$; while in the contiguous (-) monopole sector, pairs of $(PP)^{(-)1/6}$ attract and bond, forming another new type of unit, the electric W- boson, $(PP/PP)^{(-)1/3}$, due to the like attracts and bonds with like principle operative in this magnetic dimension (M4).

Both of these new single, fractional charged bosons are, however, unstable and almost instantly revert back into their respective $(VV)^{(+)1/6}$ and $(PP)^{(-)1/6}$ monopoles.

Following a period of further photon expansion, together with a resultant diminution of temperature, four $(VV)^{(+)1/6}$ monopoles are now able to bond together to create still another type of new electric unit: the double W+ boson, $[(VV/VV)(VV/VV)]^{(+)2/3}$; while four $(PP)^{(-)1/6}$ monopoles also are now able to bond together in their sector of the M4 to produce the electric double W-boson, $[(PP/PP)(PP/PP)]^{(-)2/3}$.

Both of these double bosons, like their single boson predecessors, are unstable and almost instantly revert back into their four monopole components.

The primary cause of the instability in these single and double bosons, just described, is their respective fractional charge, (+)1/3, (-)1/3, (+)2/3 and (-)2/3; all of which additionally are also electric charges generated in a magnetic dimension.

Quarks also exhibit these same four types of fractional charges and, although are much longer lasting than the boson types cited above, they still must eventually merge with other quarks to reach charge levels of (+)1, (-)1 or 0, all of which are stable and long lasting in electric dimensions.

In keeping with this understanding, as the photons in their monopoles continue to further expand, which also lowers the temperature in the M4, six $(VV)^{(+)1/6}$ monopoles are then able to bond together, creating a new type of structure, the magnetic positron, $[(VV/VV)(VV/VV)(VV/VV)]^{M}$, which is stable due to it's potential (+)1 charge in an electric dimension.

Paralleling the positron formation, six $(PP)^{(-)1/6}$ monopoles also bond together in their sector of the M4, producing a new and different type of structure, the magnetic electron $[(PP/PP)(PP/PP)(PP/PP)]^{M}$, which is also stable due to it's potential (-1) charge in an electric dimension.

As the number of magnetic electrons and magnetic positrons increase in their respective sectors of the M4, their sector properties change, as well as the sectors designation as a new magnetic dimension,the M7.

M7 Dimension Reactions

The magnetic positrons, in their sector of the new M7 magnetic dimension, are at a multibillion degree K. temperature. Here, they are extremely reactive, and thereby are able to bond easily with a contiguous positron; and, in turn, with another, and another, in what can only be described as a classic polymeric type of reaction.

Here, these almost endless strands of polymeric positrons remain stable, able only to react with each other to form new and longer chains of multi-positron polymer structures.

The magnetic electrons, created within their own sector of the M7, are also extremely reactive units, in the same multibillion degree K. temperature range as are the magnetic positrons.

They follow a pathway parallel to the positrons, forming magnetic electron polymers which coalesce in a seperate area of the M7, which is contiguous to the positron polymer area. Since both of these contiguous polymer area are of opposite 1/6 charges, they are precluded from reacting with each other due to the opposites repel rule in effect in this magnetic dimension.

It is the like attracts and bonds part of the magnetic dimensional rule that allows the bonding together of the positrons into their polymer structure, as well as allowing the electrons to bond into their polymer structure; while keeping both of the polymers stable and apart in the M7 magnetic dimension, due to the opposites repel part of the rule therein.

G7, G8, M1, M4 and M7 Reactions

Although we have previously described various dimensions as if they were separate and distinct areas, they are, in fact, composed entirely and exclusively of entities, continuous and contiguous, with absolutely no empty spaces present between them.

For example, the M1 dimension is wall-to-wall V photons and P photons, each kept in their own exclusive groupings or group areas due to the opposites repel rule in effect for these entities. There is no empty space in the M1. There is no empty space in either photon type, as both are composed entirely of respective E_V-M_V and E_P-M_P units.

Interspersed among the V and P photons are respective $[V_E]$G and $[P_E]$G gravitons. There is no empty space within each graviton. No empty space precludes other forms of matter being present in the existent universe. When a $[V_E]$G, in the G7, begins to deploy it's force, due to its inherent cyclic gravity force, G8, the force is now translated into a spinning motion which causes the $[V_E]$G to behave as if it were a propeller mixer, churning up the V photon or P photon areas in which they are positioned.

The relatively immense $[P_E]G$ gravitons are also dispersed throughout the V and P photon areas; their bodies acting as if they were baffles, directing the flow of photons, as well as allowing their surface areas to act as impact zones where V photons and P photons can slam together with like photons to form $(VV)^{(+)1/6}$ and $(PP)^{(-)1/6}$ monopoles, respectively.

In the M4, the two areas which form therein, each holding the respective $(VV)^{(+)1/6}$ and $(PP)^{(-)1/6}$ monopoles, also contain both $[V_E]G$ and $[P_E]G$ gravitons, and these then repeat the same respective spinning and baffle type actions as they did in the M1 with the V and P photons. In the case of the M4, these actions first cause the production of respective $(VV/VV)^{(+)1/3}$ and $(PP/PP)^{(-)1/3}$ bosons which, due to their fractional charge related instability, disassociate back into their respective monopole units.

The double-boson units, $[(VV/VV)(VV/VV)]^{(+)2/3}$ and $[(PP/PP)(PP/PP)]^{(-)2/3}$, which next form in the M4 due to a temperature diminution, also suffer the same fate as their respective single bosons, due to their own fractional type charges, causing them to also disassociate into respective monopoles. Their formation was produced by the same respective spinning and baffle actions of the gravitons that produced the monopoles and single bosons.

With a further diminution in M4 temperature and the same respective spinning and baffle type actions of the $[V_E]G$ and $[P_E]G$ units, six $(VV)^{(+)1/6}$ finally bond and crystalize as magnetic positrons, $[(VV/VV)(VV/VV)(VV/VV)]^M$; while in the contiguous area, the same above $[V_E]G$ and $[P_E]G$ actions cause six $(PP)^{(-)1/6}$ to bond and crystalize into magnetic electrons, $[(PP/PP)(PP/PP)(PP/PP)]^M$.

The respective spinning and baffle action of the $[V_E]G$ and $[P_E]G$ gravitons, however, does not cease with the production of magnetic positrons and electrons. Their continued action promotes the interplay of the magnetic positrons into forming polymer chains; as well as causing the newly formed magnetic electrons to produce their own polymer chains.

These exclusive areas of polymeric positrons, contiguous to

areas of polymeric electrons, together form a new magnetic dimension, the M7.

Although each of these two types of polymers appears to be a structure composed of respective magnetic positrons or electrons, the further intense internal interaction between the positrons and between the electrons in building each of their own polymer type chains causes each of the magnetic positron and electron polymers to decouple into respective (+) 1/6 and (-) 1/6 monopole groups of six, thereby producing what are really long-chain polymers of each of their respective monopoles, which are completely stable in their M7 magnetic dimension.

A key factor here is the multibillion degree K. temperature which aids in converting the complex polymer structures into monopoles and holding them as such until a further significant temperature diminution occurs.

Then, groups of six (+)1/6 monopoles will crystalize into (+) 1 positrons; and groups of six (-)1/6 monopoles into (-)1 electrons; producing an E1 electrical dimensional area, in which re-polymerization is precluded, due to the like repels rule in an electric dimension.

M7 and E1 Interactions

As the conversion of E_V and E_P, non-spatial units, into their respective spatial M_V and M_P units, in their respective V photon and P photon components of the polymerized positron and the polymerized electron structures in the M7 continues unabated, the resulting expansion in both polymerized structures causes a continuous temperature decline in both, as well as the entire M7 dimension holding them.

At the edge of each of these the two areas in the M7, each holding one of the two polymerized structures, diminution of temperature is greatly accelerated due to their great distance from the hot center. This leads to a still lower temperature range here, where depolymerization can then take place, producing respective single, free (+) 1 positrons and single, free (-) 1 electron units.

These two new electric type units will keep accumulating until a sufficient number are present to change the character of the M7 dimensional area housing them from magnetic nto an electric type of dimension, the E1.

After this transformation, the newly created (+) 1 positrons and (-) 1 electrons will then react explosively with each other, thereby creating seven new electric, reaction products, over an extended period of time, spanning four specific temperature ranges.

The $[V_E]G$ and $[P_E]G$ gravitons are also still present in the M7, as well as in the E1 dimension, as G11 gravitational entities, aiding by their spin the seven reaction processes which will be entered into by the positron and electron.

Here, in the E1, the $[V_E]G$, has transformed its cyclic gravity force into a spinning motion, which then propels the positrons and electrons to the very high speeds necessary to produce an explosive encounter between them; while the $[P_E]G$ gravitons wait passively as baffles to direct the flow of positrons and electrons into each other; as well as then directing the subsequent reaction products into a new electric area, the E4; which will form to hold each of the resultant new structures created by the two reactants, where they then stay and accumulate.

Four different temperature ranges in the E1, in declining order, cause the positron and electron reactions to produce totally different products in each of the four ranges. Following is a description of each temperature range and the reaction products which form therein.

Range No. 1

The **neutrino**, whose structure is: $[(VV/PP)(VV/PP)(VV/PP)]^0$, is the sole particle produced by the positron/electron interaction in this hottest of all the four temperature ranges. It exhibits a zero (0) electric charge.

The positron/electron reaction which created the neutrino is also the most violent of the seven positron/electron reactions produced in all of the four ranges to be presented herein.

It not only causes the breaking of each of the two

boson-to-boson bonds in each respective positron$^{(+)1}$ and electron$^{(-)1}$; each of which holds its three respective W+ bosons or three respective W- bosons together; but it also breaks the monopole-to-monopole bond in each of these respective bosons creating two (+) monopoles, $(VV)^{(+)1/6}$, and two (-) monopoles, $(PP)^{(-)1/6}$, respectively.

Thus a total of six free (+) monopoles and six free (-) monopoles thereby results from each respective positron and electron.

Each of the six free $(VV)^{(+)1/6}$, (+) monopoles from the positron, following the opposites attract and bonds rule now in effect in this new E1 electric dimension, then bonds with one of the six free $(PP)^{(-)1/6}$, (-) monopoles from the electron, creating six new Z bosons, $(VV/PP)^0$, which, in turn, then bond together into the two possible types of triads (of three Z bosons each.) Each of the two triads is a zero charged neutrino and is the 180° reverse of the other, which is depicted as follows:

1. $[(VV/PP)(VV/PP)(VV/PP)]^0$
2. $[(PP/VV)(PP/VV)(PP/VV)]^0$

In No. 1 above, each slant line (/) represents an electric monopole-to-monopole bond between two oppositely charged monopoles, a $(VV)^{(+)1/6}$ and $(PP)^{(-)1/6}$.

This is also the case in No. 2. The back-to-back)(signs in Nos. 1 and 2 indicate the presence of an electric boson-to-boson bond between a $(VV)^{(+)1/6}$ and a $(PP)^{(-)1/6}$ at the contiguous ends of two bosons. Numbers 1 and 2 are the exact same neutrino unit but reversed in space 180°.

It should be further noted that the triad structure is the basis of the positron, electron and neutrino. It will also be seen as the structural basis of the quarks, antineutrino, and gluon and antigluon, as well as the shell structure of the proton, antiproton, neutron and antineutron.

The triad structure is thus a very stable configuration in an electric dimension. In magnetic dimensions, it is only stable in a polymerized state.

Range No. 2

As the expansion of the V and P photons continues, with a concomitant temperature decline, a new temperature range is reached (Range No. 2), where insufficient energy is produced in the violent positron/electron reaction, due to the lower temperature present, to cleave each of these two reactants into six (+) monopoles and six (-) monopoles. The reaction can only produce sufficient energy to create four (+) monopoles and four (-) monopoles from the respective positron and electron.

Thus, only four Z bosons, instead of six Z bosons, are then able to be produced in this temperature range (No. 2) by the positron/electron reaction; plus one remaining intact W+ boson, $(VV/VV)^{(+)1/3}$, from the positron and one remaining intact W- boson, $(PP/PP)^{(-)1/3}$, from the electron.

The W- boson then attracts two of the four Z bosons, $(VV/PP)^0$, and bonds with them to form a new type of triad, the **down quark**, referenced as d, which structure can be depicted as follows: $[(PP/PP)(VV/PP)(VV/PP)]^{(-)1/3}$.

The W+ boson concurrently captures the remaining two Z bosons, creating a new and different triad, the **antidown quark**, $\bar{d}$ which can be depicted as follows: $[(VV/VV)(VV/VV)(PP/PP)]^{(+)1/3}$.

Also during the positron/electron reaction in temperature range No. 2, a third type of triad also was formed, the **antineutrino**.

This has been made possible through the creation of free W- bosons, $(PP/PP)^{(-)1/3}$; free W+ bosons, $(VV/VV)^{(+)1/3}$; and free Z bosons, $(VV/PP)^0$, as a result of the positron/electron reaction in temperature range No. 2, as has been previously described.

In a certain number of cases, a free W+ boson would initially bond to a free W- boson instead of a free Z boson. In the next step, the Z boson would then be added to the above pair, creating a triad made up of a W+, W- and a Z boson, which is an antineutrino; whose structure can be depicted as follows: $[(VV/VV)(PP/PP)(VV/PP)]^0$.

Upon comparison with the neutrino, $[(VV/PP)(VV/PP)(VV/PP)]^0$, one notes that the antineutrino has only one Z boson relative to the three in the neutrino. However, if one counts the number of VV, (+)

monopoles, and PP, (-) monopoles, present in each, we find that there are three of each present in both the neutrino and the antineutrino. Yet, there is a definite structural difference between the two, accounting for the different types of reactions each undergoes. However, both are primarily unreactive relative to the other four respective triads and the single type of g/ā meson, produced in the second, third and fourth temperature ranges of the E4.

As the photons in the M7 dimension continue in their unrelenting expansion, producing a further cooling of the entire dimensional area, in the peripheral areas free (+) 1 positrons and free (-) 1 electrons continue to be generated as the temperature further declines, thus eventually producing a third temperature range.

Range No. 3

In temperature range No. 3, the positron/electron reaction is still weaker, now only able to break-off a single W+ boson from the positron and a single W- boson from the electron, as well as being only able to sever the monopole-to-monopole bond in each of these two specific bosons, respectively. This is due entirely to the temperature diminution.

The two newly generated free $(VV)^{(+)1/6}$ monopoles from the W+ boson and the two newly generated free $(PP)^{(-)1/6}$ monopoles from the W- boson then interact to form two Z bosons, $(VV/PP)^{0}$.

Then, the intact remnant of the reacted positron, $[(VV/VV)(VV/VV)]^{(+)2/3}$, and one of the two Z bosons (VV/PP) just produced, then bond to create a new type of triad, the **up quark**, U, $[(VV/VV)(VV/VV)(VV/PP)^{(+)2/3}$.

Concurrently, the intact remnant of the electron from the same reaction, $[(PP/PP)(PP/PP)]^{(-)2/3}$, then captures the remaining Z boson (VV/PP) produced, creating another new type of triad, the **antiup quark**, $\overline{U}$, depicted as follows: $[(PP/PP)(PP/PP)(VV/PP)]^{(-)2/3}$.

Range No. 4

The M7 eventually further expands, with a concomitant

further decline in temperature in the peripheral areas and in the positrons and electrons therein; which then also react with each other. This reaction, however, causes only one boson to be broken off from each of the two reactants, the exact same result as had occurred in range No. 3; but, unlike the reaction in range No. 3, the severed boson is not further broken into its two component monopoles.

Thus, the positron produces the exact same remnant as in range No. 3, $[(VV/VV)(VV/VV)]^{(+)2/3}$, but the severed free W+ boson remains as $(VV/VV)^{(+)1/3}$, and not further broken down into its two component (+) monopoles, $(VV)^{(+)1/6}$ constituents.

The electron in the same reaction, simultaneously, also produces the exact same intact remnant as it did in range No. 3, $[(PP/PP)(PP/PP)]^{(-)2/3}$, and a single free W- boson, $(PP/PP)^{(-)1/3}$; which is also unlike the result in range No. 3, as it is not broken into its two (-) monopoles, $(PP)^{(-)1/6}$.

Therefore, no Z bosons form in the positron/electron reactions in range No. 4.

What, however, does form are possibly two of the most extraordinary triads yet encountered, the **gluon** and the **antigluon**.

The Organizer of the Universe

The result of the positron/electron reaction in range No. 4 was the creation of a positron remnant containing two of the three original W+ bosons, as seen in A. below:

A. $[(VV/VV)(VV/VV)]^{(+)2/3}$;

while the remnant of the electron that was formed concurrently in the reaction had only two of the three original W- bosons as its structure, seen in B. below:

B. $[(PP/PP)(PP/PP)]^{(-)2/3}$.

The W+ boson $(VV/VV)^{(+)1/3}$ broken off of the positron then instantly bonds with the above electron remnant (B) to produce a new type of triad, the **gluon** (g), illustrated as follows:

C. $[(PP/PP)(PP/PP)(VV/VV)]^{(-)1/3}$;

while the W- boson $(PP/PP)^{(-)1/3}$ detached from the electron bonds with the positron remnant (A) to create the **antigluon** (ā), depicted as follows:

D. $[(VV/VV)(VV/VV)(PP/PP)]^{(+)1/3}$.

Due to the proximity of the gluon and the antigluon when they are formed, as well as their respective (-) 1/3 and (+) 1/3 charges, an immediate bond is created between them, in accordance with the opposites attract rule in the new E4 electric dimension, thereby producing a structure of two conjoined triads, the **gluon/antigluon meson^0** or **g/ā meson**, displaying a zero (0) charge and a V-shaped configuration. This is depicted in Diagram 1, below.

The g/ā meson is a stable unit, due to its zero (0) charge. The rule in electric dimensions is that a double triad or triple triad structure is stable if it exhibits either a (0), (+) 1 or (-) 1 charge.

The g/ā meson is a conjoined double triad, which is stable and reactive at the same time. Its two conjoined triads, the gluon and antigluon, appear as two arms joined at a central focal point in a V-shaped formation, each arm with a reactive magnetic monopole at its open end.

Diagram 1

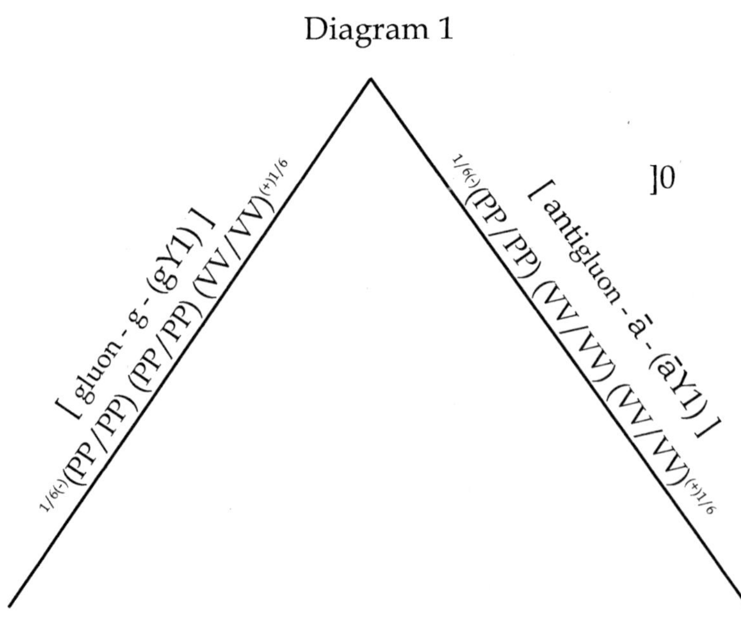

In the illustration below, the gluon, g, has a $(PP)^{(-)1/6}$ monopole, while the antigluon, ā, projected a $(VV)^{(+)1/6}$ monopole at its respective open end. Both of these monopoles are able to bond with opposite charged monopoles, which may be present; for example, at the open ends of quark type triads, such as:

1. (-) 1/3 down quarks (d)
2. (+) 2/3 up quarks (U)
3. (+) 1/3 antidown quarks ($\bar{d}$)
4. (-) 2/3 antiup quarks ($\bar{U}$)

The bond formation between a monopole at each end of the two arms of the g/ā meson and a monopole at each end of any one of the above four types of triads must follow electric dimension rules in an electric dimension.

If a bond can, therefore, be formed between the monopole at the end of each of the two arms of the g/ā meson and a monopole at each of the two ends of a quark, then the g/ā meson/quark unit which forms can then proceed to the next step; which involves adding a second quark in a continuing process which is only completed with the addition of a third quark, resulting in the creation of one of four principal entities: proton, antiproton, neutron or antineutron. This is the second of the two routes of synthesis for the above four principal entities.

If, at any one of the three quark addition steps to the g/ā meson just outlined, a bond cannot be effectuated, then the structure under construction will disassociate completely into its quark and g/ā meson components.

If the g/ā meson, depicted in Diagram 1, p.150, had first successfully attached the down quark, d, [dY1r], $[(VV/PP)(VV/PP)(PP/PP)]^{(-)1/3}$, to it's structure by appropriate bonds (+ to -), as shown in Diagram 2, (p.152), the net charge exhibited by this unit is (-) 1/3. This would be an unstable, fractional charge unit which must immediately find another suitable quark with which to bond, or disassociate back into its g/ā meson and down quark component.

Therefore, an up quark, UY1, $[(PP/VV)(VV/VV)(VV/V)]^{(+)2/3}$, was next successfully bonded to the g/ā meson/down quark unit shown in Diagram 2, p.152; resulting in the creation of a (+) 1/3, unstable intermediate two quark structure with a g/ā meson at the center, [(dY1r)(UY1) / g /ā]$^{(+)1/3}$.

This was then immediately followed by the successful bonding of a second up quark, UY1r, which resulted in the creation of a completed proton shell /g/ā meson, (UUd/g/ā)$^{(+)1}$, with a (+)1 charge; as depicted in Diagram 3, p.153, resulting from the addition of its (+) 2/3 charge to the (+) 1/3 charge of the intermediate two quark structure.

Our defining nomenclature describes the three quarks as: a (d), a (U) and another (U), as a type of triad shell (UUd)$^{(+)1}$, which, in this case, is a proton shell. A composition of a Udd^{0} would be a neutron shell; a $\overline{\text{Udd}}^{0}$ would be an antineutron shell; while a $\overline{\text{UUd}}^{(-)1}$ manifests as an antiproton shell.

If any of these four shells was produced with a g/ā meson at its center, we would add the g/ā meson to the shell name as follows: proton shell/ g/ā meson$^{(+)1}$ (UUd/g/ā)$^{(+)1}$; neutron shell/g/ā meson^{0} (Udd/g/ā)^{0}; antineutron shell/g/ā meson^{0} ($\overline{\text{Udd}}$/g/ā)^{0}; or antiproton shell/g/ā meson$^{(-)1}$ ($\overline{\text{UUd}}$/g/ā)$^{(-)1}$.

If, through a proton or neutron decay process, as we will later show, the g/ā meson is removed, the shell composed of three

Diagram 2

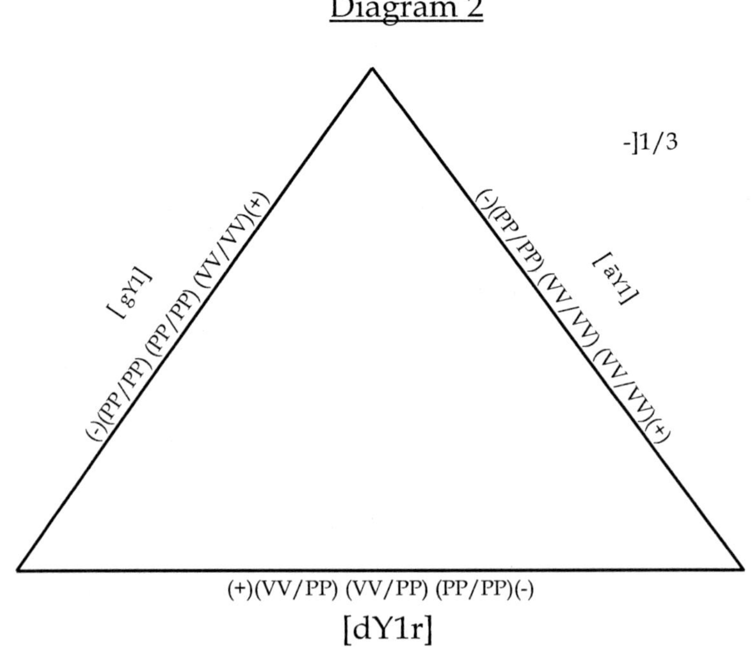

Diagram 3

Proton shell / g/ā meson $^{(+)1}$

UUd/g/ā $^{(+)1}$; [(dY1r)(UY1)(UY1r)]$^{(+)1}$

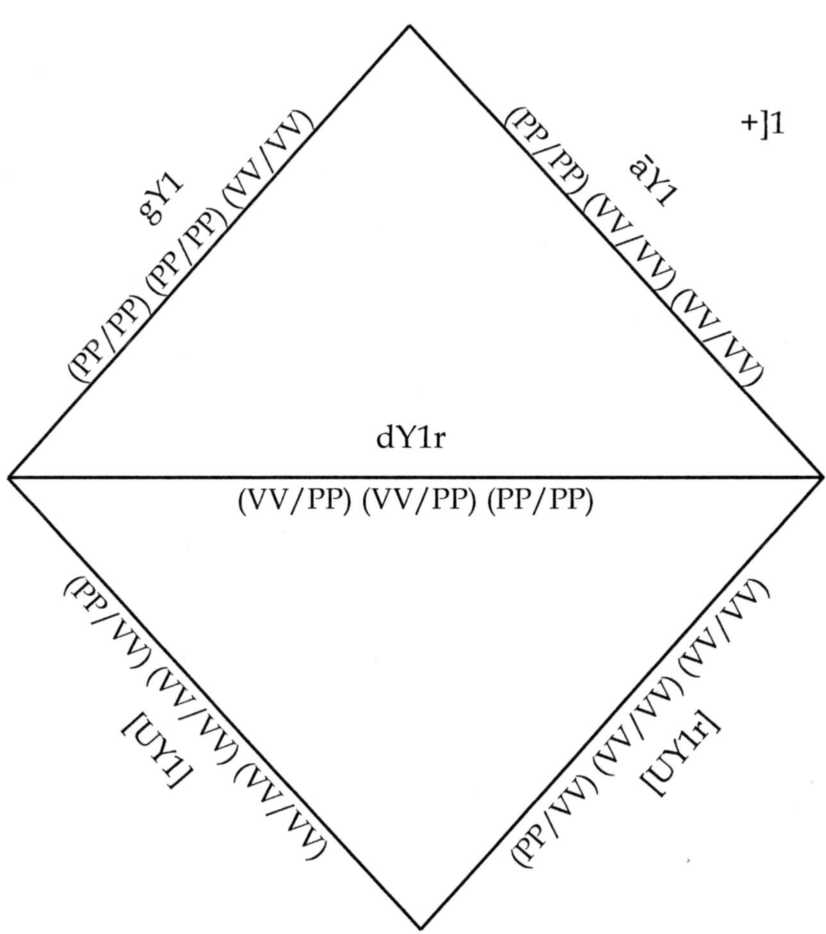

quarks will, however, survive as a stable entity: a proton shell (UUd), or a neutron shell (Udd)^{0}. The antineutron shell ($\overline{\text{Udd}}$) and the antiproton shell ($\overline{\text{UUd}}$) does not form, however, due to the conversion of all $\overline{\text{Udd}}$/g/ā/mesons and $\overline{\text{UUd}}$/g/ā mesons into protons or other particle-entities upon contact with positrons.

The type of structure drawn in Diagram 3, p.153, a (UUd/g/ā)$^{(+)1}$, is one of four ways to visually describe any of the four principal particle-entities.

Diagrams 4, 5 and 6, p.155-156, which follow, can also be used to depict a proton shell/g/ā meson$^{(+)1}$. These types of diagramed structures can also be used to depict any of the three other shell/g/ā meson entities.

The actual three-dimensional shell/g/ā meson structure is quite complex. All of the diagrams used herein are however valid approximations and used as teaching devices.

The process just described and illustrated, in fact, suffices for the organization of all four principal particle entities by the g/ā meson, and occurs exclusively in the E4 dimension.

In the E4, the proton shell/g/ā meson$^{(+)1}$ and neutron shell/g/ā meson^{0} are produced; then transferred to a new electric dimensional area, the E7; where the former eventually acquires an electron$^{(-)1}$ from the ongoing depolymerization of the electron polymers in the contiguous M11 dimension, thereby becoming an atom of hydrogen. This atom of hydrogen then forms a bond (covalent) with another same type of atom of hydrogen, producing a neutral charged molecule of hydrogen, H_2^0. These latter molecules then flow and assemble in a new neutral-electric dimension, the N1, for eventual fusion processing into the elements. These elements, then, combine into compounds in the N7. Elements and compounds are then amalgamated into the suns, planets, etc. in the final dimension, N11, of evolved existence.

g/ā Meson

The g/ā meson was the last of the seven particle-entities created in the E4 dimension, as the extant temperature slipped into it's fourth and lowest range in the E4 dimension.

(continued on p.156)

Diagram 4

Proton Shell /g/ ā meson (+)1

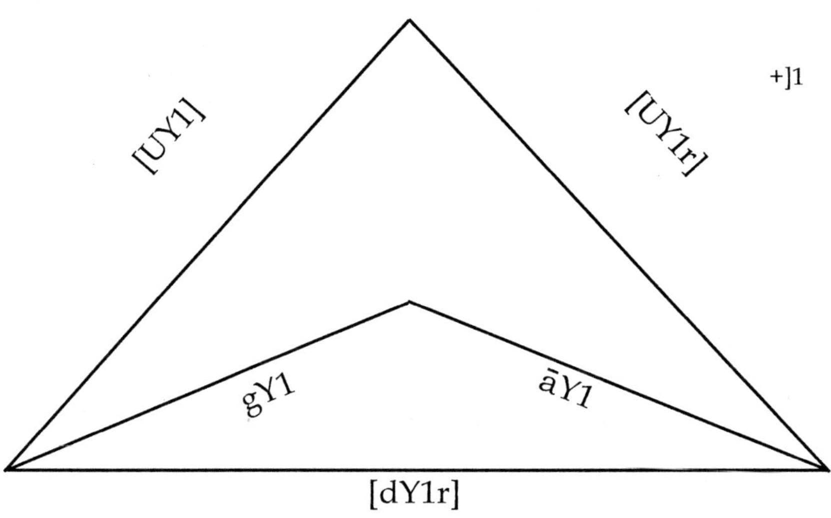

Diagram 5

Proton Shell /g/ ā meson (+)1

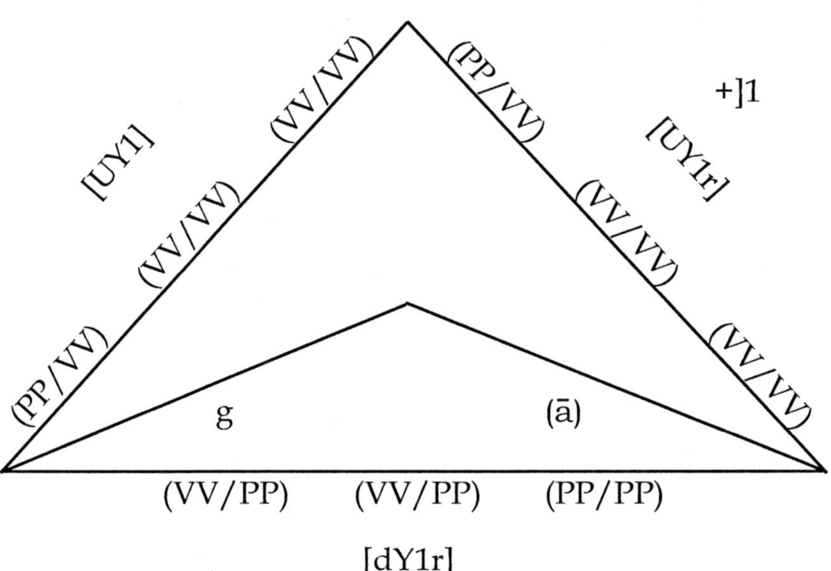

Diagram 6

Proton Shell /g/ ā meson $^{(+)1}$

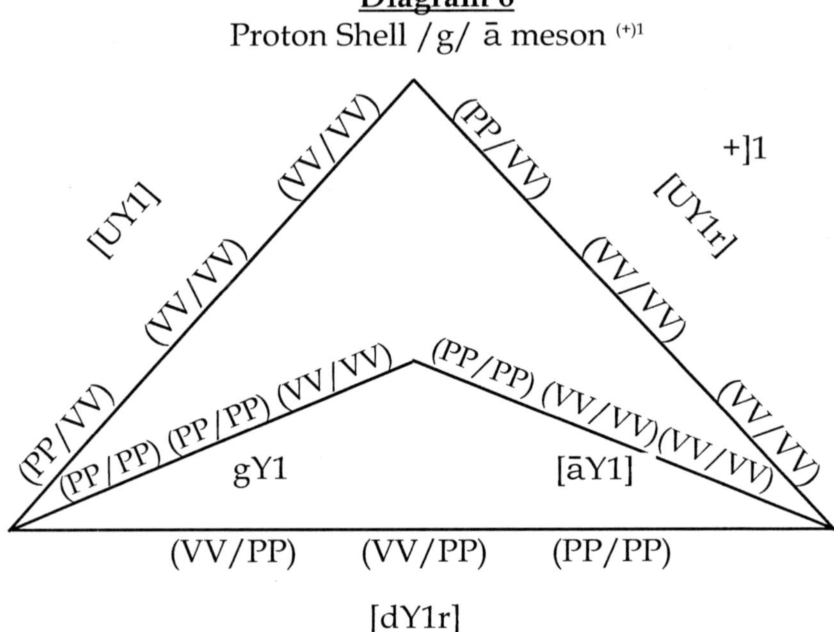

The two precursors of the meson, the gluon$^{(-)1/3}$ and the antigluon$^{(+)1/3}$, survived as independent units in the E4 for a fraction of an instant before having to bond together into the g/ā meson, a stable zero (0) charged entity.

We count the g/ā meson as a merger product and as the seventh entity, created in the E4, instead of the gluon and antigluon as the seventh and eighth entities, respectively, and the meson as the ninth. The gluon and antigluon are simply too fugitive to be counted as anything but intermediate entities.

It is the g/ā meson, however, which is the indispensable key to the creation of the four fundamental particle-entities: the proton, the neutron, the antiproton and antineutron; and without which the synthesis of the four could not occur.

Of the six triad entities produced prior to the creation of the g/a/meson, only the following four will interact with the g/ā/meson upon its entry into the E4:

1. the up quark (U)
2. the down quark (d)
3. the antiup quark ($\overline{U}$)
4. the antidown quark ($\overline{d}$)

The other two triads, the neutrino (V) and the antineutrino ($\overline{V}$), take no part in the direct synthesis of the four fundamental entities, other than acting as the reaction medium in which the above four triads and the g/ā meson interact to produce the various intermediates and the final four fundamental entities. Another type of reaction medium, water, also exhibits the same useful and parallel property in the N7 dimension in which compounds are made.

The initial reaction between the monopoles at the ends of the two arms of the g/ā meson and the monopoles at the two ends of the specific triad (up quark, down quark, etc.) produces an intermediate triple triad structure out of the double triad of the g/ā meson, thus momentarily stabilizing both:

Figure I, p.158, is a double triad; the g and ā arms, each being a triad structure.

Figure II, is the double triad in Figure I to which a (d) triad, a down quark, has been bonded, thus converting the double triad g/ā meson into a triple triad intermediate with a $^{(-)1/3}$ charge.

Any of the other previous three triads can also be attracted to and bond with a g/ā meson in a similar way. In each of these latter three cases, the final charge on the intermediate will also be fractional.

The fractional charge on each of the four intermediate structures (Figures II, III, IV and V) causes each of them to then attempt to add, in consecutive order, two additional triads in order to produce an end product which is a stable, triple triad shell structure: which, in turn, has a g/ā meson as a central nucleus. The final structure, in order to remain stable, must also exhibit either a (0), (+) 1 or (-)1 charge. Only these three types of charges are stable.

This double requirement of a triple triad structure and having one of the three types of charges cited above, is an absolute requirement for long-term stability.

For example, if the fractional charged structure in Figure II below, were able to bond with a second quark triad, a U(+)2/3 up quark, an intermediate structure, as seen in Figure VI, p.159,) is produced, projecting a fractional (+) 1/3 charge.

Immediately, another quark must be added in order to:

1) convert the double, d/U/g/ā, structure into a triple triad;
2) convert the (+) 1/3 charge into one of the three stable charges (0, +1, -1).

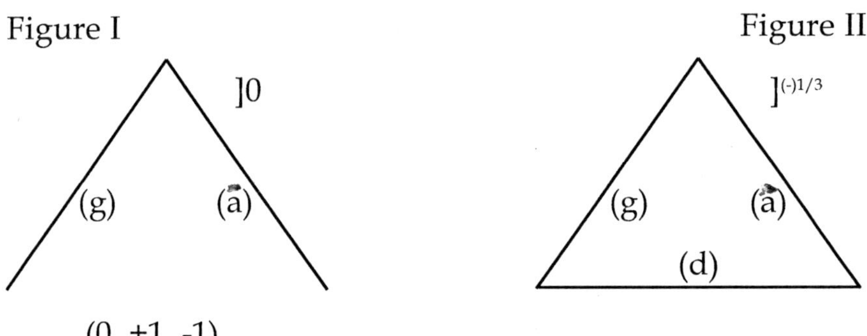

Two quark types will satisfy these conditions: 1) the U(+)2/3 and 2) the d(-)1/3. The former would produce a UUd/g/$\bar{a}^{(+)1}$, the proton/ g/ā meson with a (+) 1 charge, while the latter would produce a Udd/g/$\bar{a}^{0}$, the neutron/g/ā meson, with a zero (0) charge. Each type is respectively depicted in Figures VII and VIII,

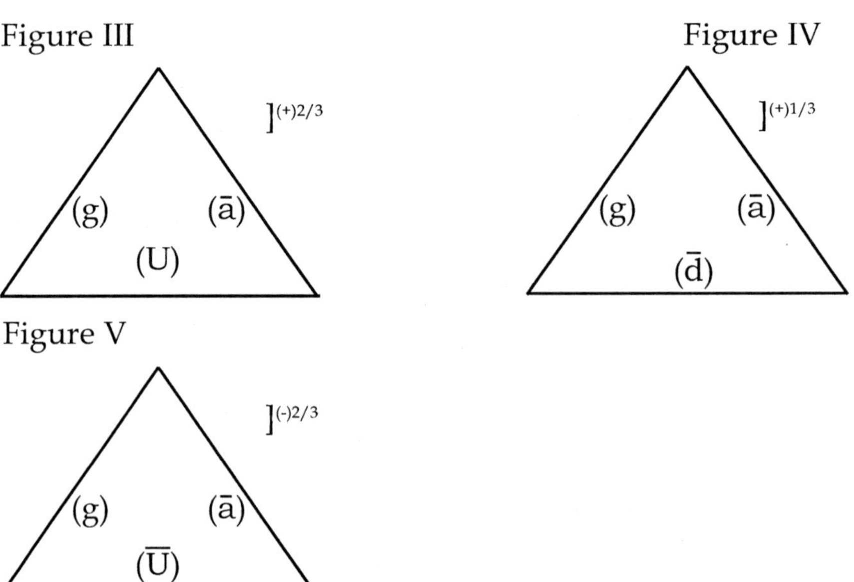

p.160.

On the other hand, if the first quark to bond with the g/ā meson were an antidown quark, $\bar{d}^{(+)1/3}$, instead of the $d^{(-)1/3}$, down quark, then only the addition of two $\bar{U}^{(-)2/3}$, antiup quarks, or one $\bar{U}^{(-)2/3}$ antiup quark and one $\bar{d}^{(+)1/3}$ antidown quark, could respectively form a stable $\overline{UUd}$/g/$\bar{a}^{(-1)}$, antiproton/g/ā meson$^{(-)1}$; or, in the second case addition, a $\overline{Udd}$/g/$\bar{a}^{0}$, antineutron/g/$\bar{a}^{(-1)}$ meson^{0}, each also being a whole number (-) 1 or 0, respectively. See Figures IX and X, p.160.

The substitution of a single $U^{(+)2/3}$ quark or a $d^{(-)1/3}$ quark in either of the two structures, as depicted in Figures IX and X, for one of the antiup or antidown quarks, respectively, would result in an unstable fractional charge structure (see Figures XI and XII), both of which would instantly disassociate into their three quarks and

Figure VI

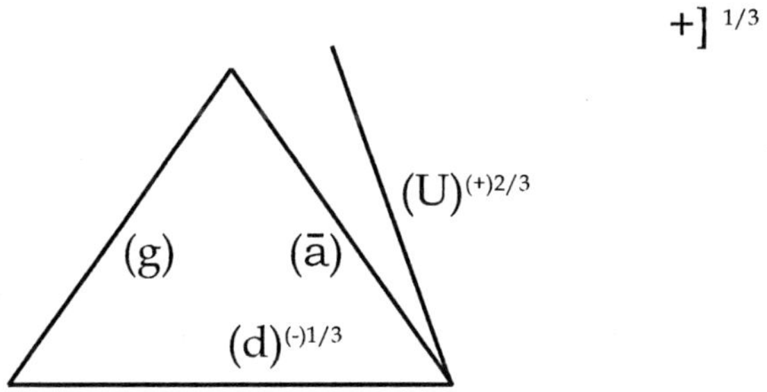

g/ā meson components.

Thus, the stability requirements of a triple triad shell structure (with a g/ā meson at its center), exhibiting a whole number charge ((+) 1, (-) 1, 0), precludes the establishment of all other structural type particle-entities in the E4 dimension except: the proton shell/g/ā meson$^{(+)1}$, the neutron shell/g/ā meson^{0}, the antiproton shell/g/ā meson$^{(-)1}$ and the antineutron shell/g/$\bar{a}^{0}$ meson. Only these four stable particle-entities can survive, and then accumulate in the E7 dimension for further reaction processing.

Examples of precluded entities can be seen in Figures XIII through XVIII, p.161. All of these structures are fractional charged, thereby violating the whole number charge rule for stability and

Figure VII: $UUd/g/\bar{a}^{(+)1}$
Proton Shell /g/a/meson $^{(+)1}$

Figure VIII: $UUd/g/\bar{a}^{0}$
Neutron Shell /g/a/meson ^{0}

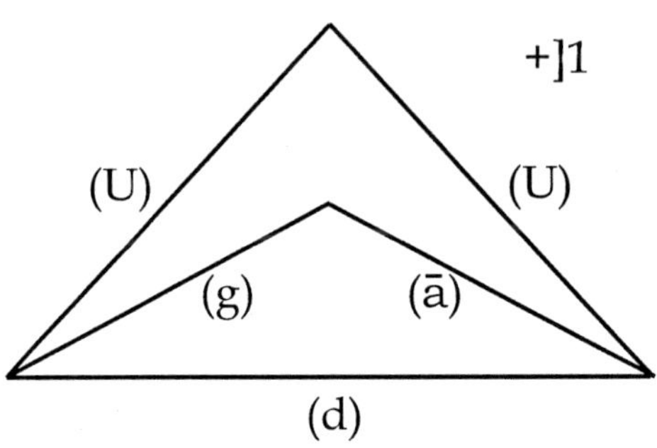

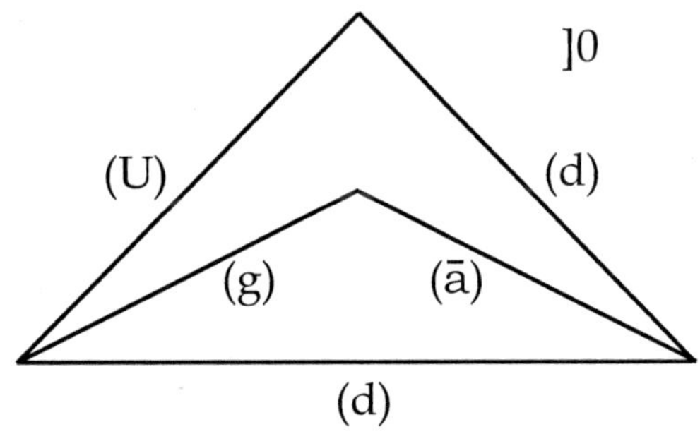

Figure IX: $\overline{UUd}/g/\bar{a}^{(-)1}$
Antiproton shell /g/a/meson (-)1

Figure X: $\overline{Udd}/g/\bar{a}^{0}$
Antineutron shell /g/a/meson ^{0}

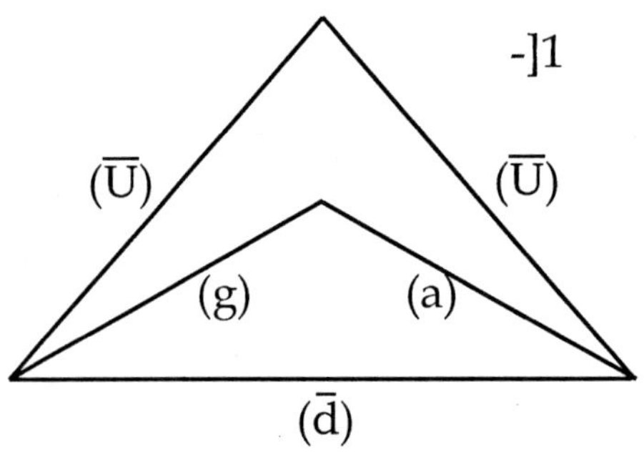

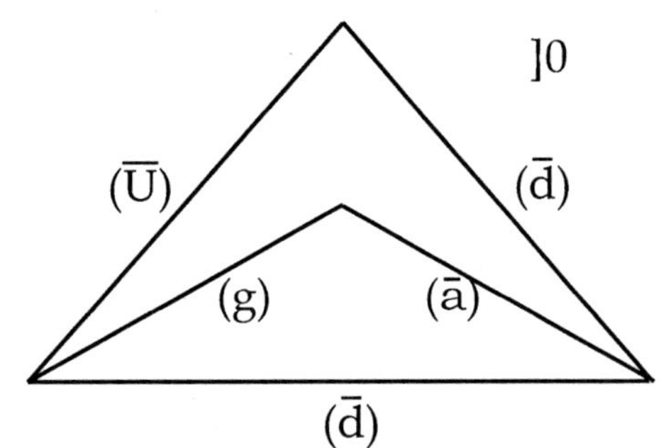

Figure XI: $U\overline{Ud}/g/\bar{a}^{(+)1/3}$

Figure XII: $\overline{U}dd/g/\bar{a}^{(-)2/3}$

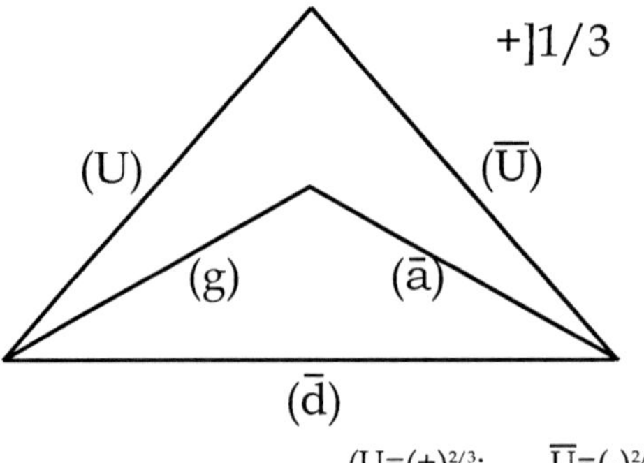

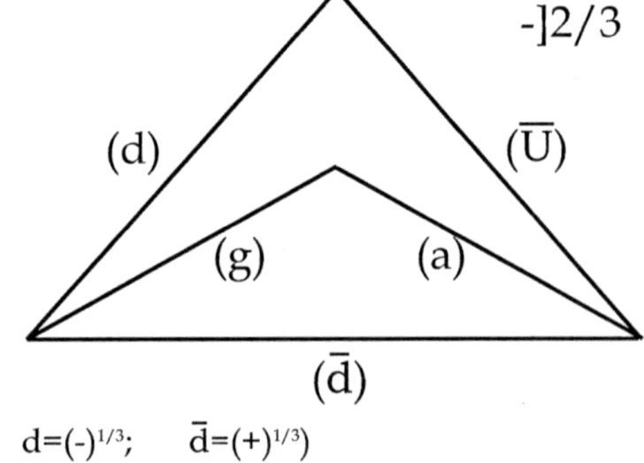

$(U=(+)^{2/3}; \quad \overline{U}=(-)^{2/3}; \quad d=(-)^{1/3}; \quad \bar{d}=(+)^{1/3})$

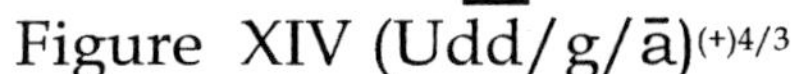

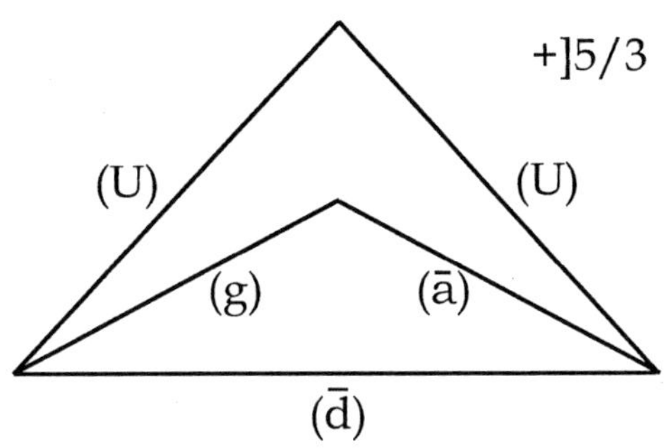

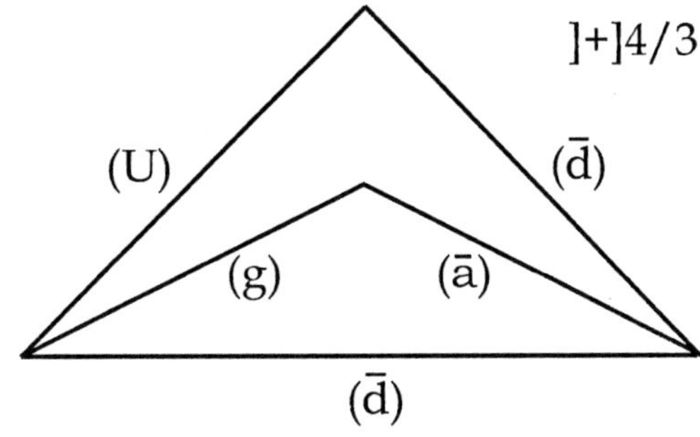

Figure XV (Udd̄/g/ā)$^{(+)2/3}$

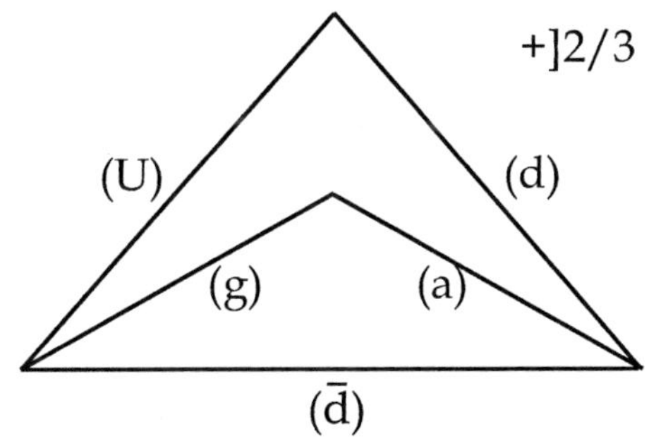

Figure XVI (Ūdd/g/ā)$^{(-)4/3}$

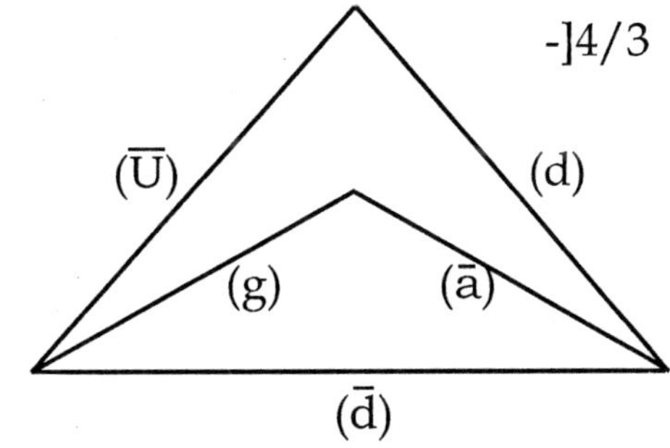

Figure XVII (Udd̄/g/ā)$^{(+)2/3}$

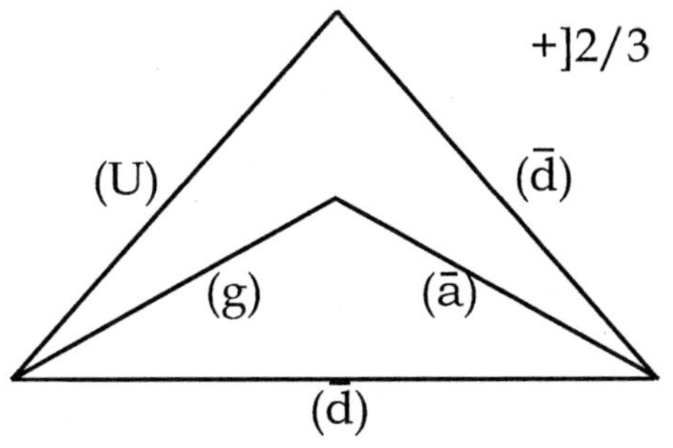

Figure XVIII (ŪŪd/g/ā)$^{(-)5/3}$

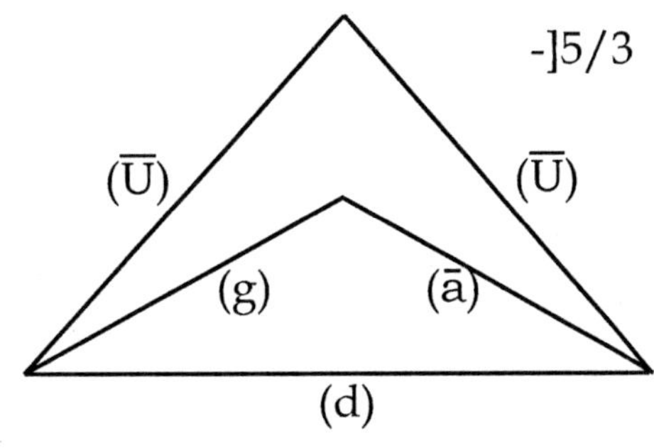

thus, cannot survive, and must disassociate into their components.

Quark Composition Rules

The rules governing the composition of particle-entities containing various quarks have been illustrated. The following rules govern the creation of any particle-entity with mixed quark components: up and down quarks combined with an antiup or antidown quark:

1. No up quark (U) or down quark (d) can be bonded to a g/ā meson which has initially bonded to an antiup quark ($\bar{U}$) or an antidown quark ($\bar{d}$).
2. No antiup quark ($\bar{U}$) or antidown quark ($\bar{d}$) can be bonded to a g/ā meson which has initially bonded to an up quark (U) or down quark (d).

Violation of the above rules will result in an entity that has a fractional charge: (+)2/3, (-)2/3, (+)5/3, (-)5/3, (+)4/3, (-)4/3.

All such fractional charge units are unstable and will disassociate back into their component quarks and g/ā meson.

Other Unstable Quark Combinations

There are other types of quark and g/ā meson combinations which may form, that are also unstable and quickly disassociate into their quark and g/ā meson components.

For example: $ddd/g/\bar{a}^{(-)1}$ and $\bar{d}\bar{d}\bar{d}/g/\bar{a}^{(+)1}$ are both conjoined triple triads, with a g/ā meson at its center, which respectively exhibit (-)1 and (+)1 whole number charges. Both seem to fit the required criteria for a stable, fundamental particle, except that both must obey the overriding rule in electric dimensions that like particles repel. Since each contains only like d or $\bar{d}$ quarks, respectively, as their triple triad structures surrounding the central meson, each of the three like quarks would repel each other, causing them to explosively disassociate into their quark and meson components. $UUU/g/\bar{a}^{(+)2}$ and $\bar{U}\bar{U}\bar{U}/g/\bar{a}^{(-)2}$ would also disassociate for the same reason.

Under certain unusual conditions, the gluon (g) or the antigluon (ā) can, for an instant, become part of a shell in a shell/g/ā/meson combination, such as $Udg/g/\bar{a}^{0}$ or

Udā/g/ā$^{(+)2/3}$. These two structures also result in disassociation into their components. Here, it is due to the unusual structure of the gluon and antigluon, which have no Z bosons present in their structure. This causes a structural anomaly which does not permit either (g) or (ā) to bond with quarks to form a triad structure. They can only bond with each other to become a g/ā meson. The meson, in turn, can only bond with a three quark structure. The g or ā type of combination with a quark cannot form as an particle-entity and is thus removed as a player in all of these reactions.

Finally, there are the "shell" structures, UUd or Udd, derived directly as a result of certain types of reactions, previously discussed, in which the g/ā/meson has been completely removed.

The shell proton$^{(+)1}$, UUd$^{(+)1}$, and the shell neutron^{0}, Udd^{0}, are the exact same outer structures which formerly surrounded the g/ā/meson in the respective UUd/g/ā$^{(+)1}$ and Udd/g/ā^{0} units.

The shell units are as stable as the shell g/ā/meson unit structures, except that under extreme energy fluxes produced in certain reactions, or in conditions in which they are placed under stress by being twisted at an extreme angle relative to their normal spatial configuration, a double-quark bond structure between a shell proton (UUd)* and another adjacent shell proton**, which attaches and fuses both of them together, along with everything connected to each, creating a complex chain of protons and neutrons in an element such as ${}_{92}U^{234}$, can thus be completely severed.

This splitting action then allows each of these two formerly

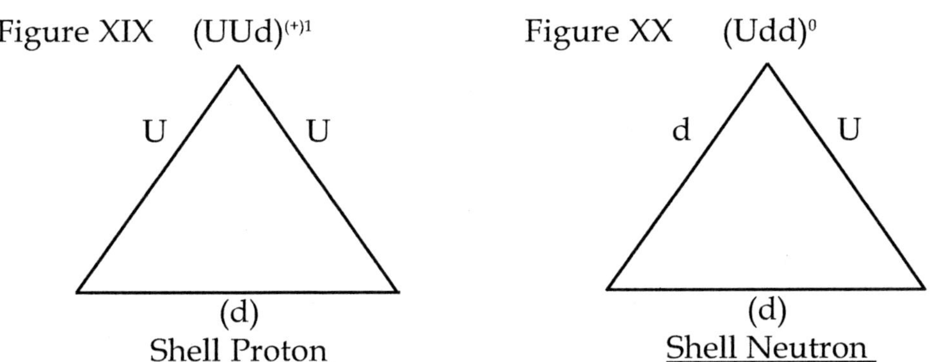

* Can also be a shell neutron (Udd). See figures XIX and XX

** Can also be a shell neutron, or a shell neutron/g/ā meson, or a shell proton/g/ā meson. See figures XIX and XX

bonded and fused proton shells (UUd) to each then become a unit in each of the two resultant separate and distinct new elements, ${}_{90}Th^{230}$ and $({}_{2}He^{4})$ which form. This is shown in the following rendering of what has occurred to ${}_{92}U^{234}$ as a result of this type of splitting.

$$ {}_{92}U^{234} \rightarrow \quad {}_{90}Th^{230} \quad + \quad ({}_{2}He^{4})^{(+)2} $$

One of these two proton shells, which was severed from the other, became part of the ionized helium nucleus (alpha particle). It was bonded to a neutron shell, which, in turn, was bonded to another proton shell, which was, in turn, attached to another neutron shell. This alpha particle, carries a (+) 2 charge. The other formerly bonded proton shell remained attached to the remaining fraction of the ${}_{92}U^{234}$ which becomes the severed ${}_{90}Th^{230}$ unit.

This is the mechanism responsible for the production of alpha particles $({}_{2}He^{4})^{(+)2}$ and various other types of spontaneous fission products .

Only shell proton$^{(+)1}$ ($UUd^{(+)1)}$) and shell neutron^{0} (Udd^{0}) structures have been produced; shell antiproton$^{(-)1}$ ($\bar{U}\bar{U}d^{(-)1}$) and shell antineutron^{0} ($\bar{U}dd^{0}$) structures were never created due to all of the antiproton/g/ā meson$^{(-)1}$ and antineutron/g/ā meson^{0} structures extant having been reacted with positrons, and, thereby, have been converted into proton/ g/ā mesons, as well as other particle-entities, such as; electrons, positrons, neutrinos and anti neutrinos.

Dimension Categories

Each of the eleven dimensions composing each of the first four phases (G, M, E, N) is categorized as being one of three types:

I. Entity, II. Force or III. Field.

I. The entity can be: a wave type (V); a particle type (P); or a mixed type,containing V and P.

In Chart XX, p. 117, which illustrates the entire 55 dimensional system, all of the entity types are found in the middle column, or middle row of dimensions, in each of the first four phases; specifically in #1, #4, #7, #10 and #11 dimensions of each phase, as seen below:

Phase	Entity Dimensions
G	G1, G4, G7, G10, G11
M	M1, M4, M7, M10, M11
E	E1, E4, E7, E10, E11
N	N1, N4, N7, N10, N11

The V and P type particle-entities inhabiting the central column of each of the four phases are as follows:

Phase	Entity
G	$[V_E P_E]G$, $[V_E]G$, $[P_E]G$
M	$V^{(+)1/12}$, $P^{(-)1/12}$, $(VV)^{(+)1/6}$, $(PP)^{(-)1/6}$, positron polymer, electron polymer
E	$(VV/VV)^{(+)1/3}$(W+) boson; $(PP/PP)^{(-)1/3}$,(W-) boson; $(VV/PP)^{0,}$ (Z) boson $(VV/VV)(VV/VV)(VV/VV)]^{(+)1}$ (positron) $[(PP/PP)(PP/PP)(PP/PP)]^{(-)1}$ (electron) $[(VV/PP)(VV/PP)(VV/PP)]^{0}$ (neutrino) $[(VV/PP)(VV/PP)(VV/VV)]^{(+)1/3}$ (antidown quark) $[(VV/PP)(VV/PP)(PP/PP)]^{(-)1/3}$ (down quark) $[(VV/PP)(VV/VV)(VV/VV)]^{(+)2/3}$ (up quark) $[(VV/PP)(PP/PP)(PP/PP)]^{(-)2/3}$ (antiup quark) $[(VV/PP)(VV/VV)(PP/PP)]^{0}$ (antineutrino) $[(VV/VV)(PP/PP)(PP/PP)]^{(-)1/3}$ (gluon) $[(PP/PP)(VV/VV)(VV/VV)]^{(+)1/3}$ (antigluon) $[(VV/VV)(PP/PP)(PP/PP)(VV/VV)(VV/VV)(PP/PP)]^{0}$ (g/ā meson) Proton shell/g/ā meson$^{(+)1}$, neutron shell/g/ā meson^{0}, antiproton shell/g/ā meson, $^{(-)1}$,antineutron shell/g/ā^{0}

$\underline{N}$ $\,_{2}meson^{0}, ({}_{1}H^{2})^{(+)1}, ({}_{1}H^{3})^{(+)1}$
$^{2}({}_{1}H^{1}), {}^{2}({}_{1}H^{2}), {}^{2}({}_{1}H^{3})$

The elements of the periodic table.
The compounds composed of the elements of the periodic table.

The ferromagnetic element cores of planets.
The suns and planets of the existent universe.

II. A force is present and is being exerted in each V type entity in the #2, #5 or #8 dimension of each of the first four phases; these three force dimensions comprise the right-hand column of each of these four phases. The gravitational phase's force is the gravity force; the magnetic phase has the magnetic force; and the electric phase has the electric force. Each of the latter two forces, as previously discussed, is a minute expression, or a part of the gravity force.

The neutral-electric phase's V type entities also exercise the electric force, but it is neutralized by the electric field produced by the P type entities in their own field dimensions, therein.

Phase	Force Dimensions
G	G2, G5, G8
M	M2, M5, M8
E	E2, E5, E8
N	N2, N5, N8

Each type of force is expressed exclusively by a wave type entity in each of the force dimensions above.

In the gravitational phase (G), the gravity force is expressed by the $[V_E P_E]$ graviton and the $[V_E]$ graviton ($[V_E]G$); while in the magnetic phase the V photon, $(V)^{(+)1/12}$; and the (+) monopole, $(VV)^{(+)1/6}$ are the entities expressing the magnetic force. The electric phase has its electric force projected by the W+ boson $(VV/VV)^{(+)1/3}$, the positron $[(VV/VV)(VV/VV)(VV/VV)]^{(+)1}$, the proton shell/g/ā meson$^{(+)1}$ (UUd/g/ā)$^{(+)1}$ and the proton shell$^{(+)1}$ (UUd)$^{(+)1}$.

An entity exercising the electric force in the neutral-electric phase is the proton shell/g/ā meson$^{(+)1}$, which is composed of thirty-six V photons, each at (+) 1/12, and twenty-four P photons at (-) 1/12; an excess of twelve V photons with an expressed charge of (+) 1, or an expansive force of (+) 1. Later, we will discuss the proton shell$^{(+)1}$ as an electric force expressing entity and its place in the structure of the isotopes of all of the elements.

These latter two entities are each neutralized by an electron$^{(-)1}$, which is a P type of electrical field entity, which does not express a force, but fields an immense body, relative to the size of the body of a V type entity. It can resist the expansion force of the proton shell/g/ā meson$^{(+)1}$, or proton shell $^{(+)1}$, without expressing a counterforce. Thus, the electron$^{(-)1}$ resistance to an expansion force of (+) 1 is (-) 1, bringing the proton shell/g/ā meson$^{(+)1}$, or proton shell $^{(+)1}$, to a neutral equilibrium status, called an atom of hydrogen, $({}_1H^1)^0$

III. Field. The gravity, magnetic, electric and neutral-electric fields are found exclusively in the three dimensions of the left-hand column of the G, M, E and N phases, specifically the #3, #6 and #9 dimensions of each phase, as follows:

Phase	Field Dimensions
G	G3, G6, G9
M	M3, M6, M9
E	E3, E6, E9
N	N3, N6, N9

Each P type entity expresses a resistance to the expansion of a V type entity. This resistance, in the form of increasing rigidity of the P type body as the expansive force increases, is the expression of a field. It is not a counterforce but a field of resistance.

In the gravitational phase it is expressed by the $[V_EP_E]$ graviton and the $[P_E]$G graviton. In the magnetic phase it is the P photon $(P^{(-)1/12})$ and the (-) monopole, $(PP)^{(-)1/6}$, while in the electric phase, it is the W-boson and the electron$^{(-)1}$.

Charges: V Photon and P Photon

The charge exhibited by the V photon (V) is 1/12 of the (+) 1 charge produced by a positron or a proton. This (+) 1/12 charge, as well as the (-)1/12 charge of the P photon and all others at (+) or (-) 1/6, are categorized as magnetic. All charges above this threshold are electric.

The $V^{(+)1/2}$ photon charge is the sum total of all of the minuscule (+) type charges produced by each of the myriad number of E_V and M_V type subunits present within it; each of these subunits have the exact same minuscule (+) type charge.

Therefore, since each E_V subunit has the same (+) charge as each type of M_V subunit (M_V^F , M_V^S), and as the combined number of E_V and M_V subunits in each V photon extant is always the same, therefore the charge on every V photon in the existent universe will always be (+) 1/12.

The charge displayed by a P photon (P), however, is 1/12 of the (-) 1 charge demonstrated by an electron, or (-) 1/12.

The P photon charge is the sum total of all of the minuscule (-) charges exhibited by each of the myriad number of E_P and M_P type subunits present within it; both subunits having the exact same charge.

In parallel to the V photon, since each E_P subunit has the same charge as each type of M_P subunit (M_P^F , M_P^S), and the combined number of both types of subunits in each P photon extant is always the same, the charge on every P photon extant in the existent universe will always be (-) 1/12.

The (+) 1/12 generated by the V photon and the (-) 1/12 of the P photon result from the physical separation caused by the severing of each from their respective parent graviton.

Thus, each of the (+) or (-) 1/12 charges is an <u>impulse attraction property</u> in which each respective photon expresses its need to once again reunite with a progenitor graviton in order to achieve the "1" or unified state.

Even though the conditions needed for this reunification with their respective gravitons are no longer extant, the respective photons in the M1 magnetic dimension still express the impulse but transform it into an action which unites like photons in each of their homogeneous halves of the M1.This results in the production of respective double-photon structures, the (+) and (-) monopoles: $(VV)^{(+)1/6}$ and $(PP)^{(-)1/6}$, respectively.

This impulse attraction principle becomes the underlying cause for the organization of the photon universe.

The impulse property to reunite is also still a part of the unification impulse that is present in all gravitons as well.

When the $[V_E]$G expands, it attempts to compress the $[P_E]$G, which responds by invoking an ever- increasing gravitational field response, expressed as rigidity in its substance, a passive invocation.

When the $[V_E]$G finally deflates, going from it's maximum apex gravity force to the zero gravity force of its sine wave function, it pulls the contiguous $[P_E]$G into inflating its size to the size which the $[V_E]$G had maximally expanded to, at the apex of its sine wave function. The pull is an impulse attraction property to reunite the $[V_E]$G and $[P_E]$G.

The photons, V and P, which are produced in each type of respective graviton from their respective substance, translate and keep the attraction impulse (charge) of their $[V_E]$G and $[P_E]$G progenitors, but in relatively minuscule units (+ or - 1/12), in keeping with their relative minuscule percentage of the massive substance of their respective gravitons.

The impulse to once again reunify takes the form of a universal attraction between all of the entities in the existent universe in order to once again repeat the partition event, but in reverse, and once again merge into and become the Singularity.

Phase and Dimension Generation

In each of the first four phases (G, M, E, N), the entities in each of their first dimensions (#1) then evolve, through various actions and reactions, through an additional series of ten dimensions, for a total of eleven dimensions for each of the four phases.

In each of the first three phases (G, M, E), the evolving entities eventually enter the seventh dimension of each phase in which they then create new types of entities, which then move away from their creator entity's seventh dimension and originating phase, in order to establish a new type of dimension which is always the first dimension, of a new phase.

For example, the $[V_E]G$ and $[P_E]G$ gravitons, together as the $[V_EP_E]G$ merged entity in the first dimension of the gravitational phase, designated as G1, are transformed by various processes in play in each of the five successive dimensions, G2 through G6.

Then, in the G7 dimension of the gravitational phase, the $[V_EP_E]$ G gravitons finally separate completely into their component $[V_E]G$ and $[P_E]G$ gravitons. The process that led to this separation took place mainly in the G4 dimension. During that period, an internal process caused each of the two different merged gravitons to create from their own substance new entities: the V photons ($V^{(+)1/12}$) and P photons ($P^{(-)1/12}$), respectively. These were then completely disgorged by each of the respective gravitons in the G7 dimensions, resulting, in turn, in the $[V_E]G$ and $[P_E]G$ achieving complete separation and individuation as G7 entities.

Each of the newly created V and P photon types then migrate and assemble in their own separate and segregated areas of a new dimension, the M1, which is also the first dimension of a new magnetic phase.

The $[V_E]G$ and $[P_E]G$ gravitons are, as a consequence, dispersed throughout both V and P areas of the M1, as G7 entities.

Essentially, each of their distinct gravitational $[V_E]G$ and $[P_E]G$ bodies, even though separated from each other, is its own circumscribed G7 area, a separate and complete G7 dimension.

As a new cyclic force of gravity wave (G8) is generated in each $[V_E]G$, its expansion is now translated, however, into a spinning motion, causing the V photons and P photons among which they are interspersed, to be propelled to high speeds and violent collisions with like V photons and like P photons, respectively.

These collisions take place on the $[P_E]G$ graviton bodies which act as anvils upon which the propelling, hammer-like $[V_E]G$ gravitons slam together pairs of like photons, producing two new entities: the (+) monopole, $(VV)^{(+)1/6}$, and the (-) monopole, $(PP)^{(-)1/6}$.

When the $[V_E]G$ generated it's new cyclic gravity wave and spinning motion, it is considered to be in a new dimension, the G8, one of the three gravitational force dimensions. The other two gravitational force dimensions are the G2, in which the gravity force sets up the structure and the initial size of the $[V_E]G$; and G5, in which the gravity force further expands the $[V_E]$G and causes it to produce V photons from its substance.

The $[P_E]$G, when its surface and body act as an anvil, as well as a baffle in directing the flow of new monopoles forged on its surface, is considered to be a G9 gravitational field. This is one of three gravitational fields. The other two are the G3, in which the $[P_E]$G first generates a resistance type response to the expansive intrusion of the $[V_E]$G, as well as its initial size; and G5, in which the $[P_E]$G reacts to an attempted compression by the $[V_E]$G , causing it to produce a rigidity in its substance, which upon being pulled into its expansive mode, produces P photons.

The Organizer Mechanism

The g/ ā meson is a reactive nucleus with two arms projecting outward from a common base in a V-shaped pattern. One arm is a (-)1/3 charged gluon, g, [(PP/PP)(PP/PP)(VV/VV)]; while the other arm is a (+)1/3 charged antigluon, ā, [(VV/VV)(VV/VV)(PP/PP)], as seen in Diagram 7, p.172.

One may view this structure and its actions as two fishing poles, each baited with a same or different charge at its end, catching opposite charged quarks (up or down, antiup or antidown), leading to the construction of a three quark shell about the g/ā meson. These shells are the proton$^{(+)1}$, neutron^{0}, antiproton$^{(-)1}$ or antineutron^{0} ($UUd^{(+)1}$, Udd^{0}, $\overline{UUd}^{(-)1}$ and $\overline{Udd}^{0}$), respectively.

As a result, the g/ā meson also becomes the internal reinforcing and strengthening framework of these quark shells, much like the steel framework upon which a building is constructed.

The Work of the Organizer Illustrated

In the course of the following exposition to illustrate another way of viewing the working mechanism used to construct the proton, neutron, antiproton and antineutron, we will use two-dimensional depictions for teaching and explanation.

As the actual g/ā meson, as well as the quark structures, are

Diagram 7

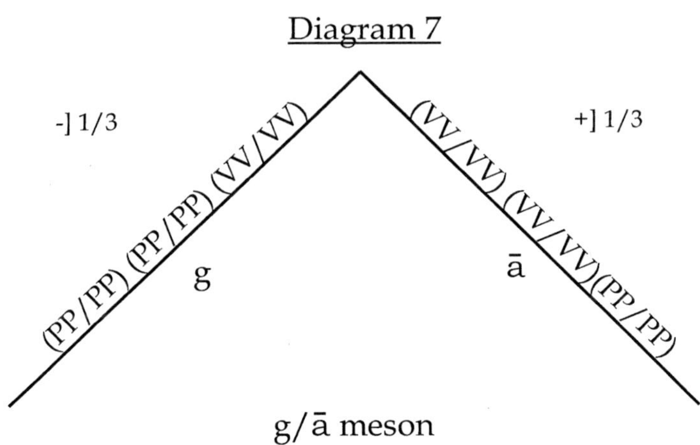

quite complex three-dimensional constructs, the reader is advised that the illustrations and the picturesque language of visualization used are for teaching purposes and thus limited. They are, however, good approximations of what occurs.

Two factors direct the additions of the four types of quarks (d, $\bar{d}$, U, $\overline{U}$) to bond together around the g/ā/meson.

The first factor is the presence of a (-)1/3 or (+)1/3 charge on the respective g and ā arms of the meson. These charges attract opposite charges on the quarks, while repelling like charges on quarks, in keeping with the rules of attraction and bonding in the E4 electric dimension.

The second factor relates to the topographical features of the quark structures. Due to the great complexity of this second factor, we will attempt to use simple analogs of these structures in order to illustrate the pertinent factors that determine the success or failure of bonding in which only four stable fundamental structures can possibly be produced: the proton shell/ g/a meson$^{(+)1}$, the neutron shell/g/ā meson^{0}, the antiproton shell/ g/a meson$^{(-)1}$ and the neutron shell/g/ā meson^{0}.

d and $\bar{d}$ Quarks

The structure of the d and the $\bar{d}$ quark can be depicted as a simple line with an arrow-key at each end, plus a charge designation at its midpoint to indicate type and quantity of charge.

Example 1 illustrates a d quark, while Example 2 illustrates a $\bar{d}$ quark.

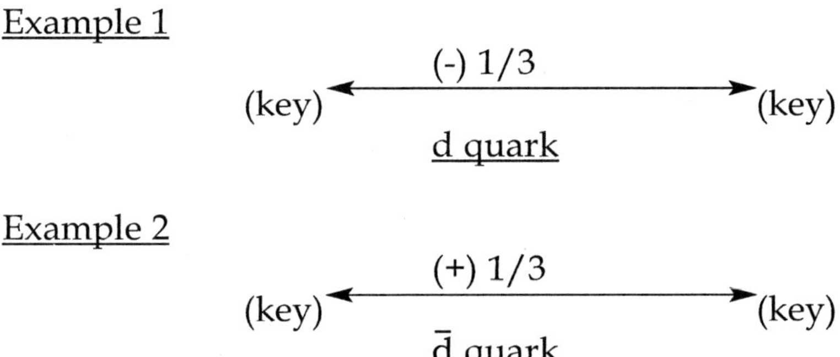

The arrow key (→) on each end of the above two examples can only penetrate and fit into lock-like circle structure, ○, which is present on both ends of U quarks and $\overline{U}$ quarks, as depicted in respective Examples 3 and 4, below.

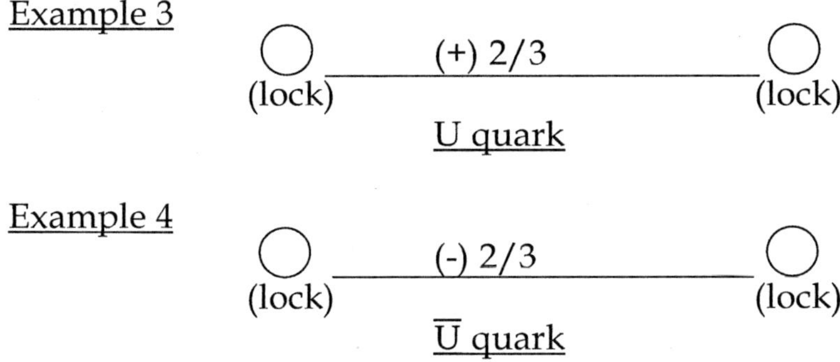

The rules guiding the interaction of the above four types of quarks are:

A. Arrow keys, on d and $\bar{d}$ quarks, cannot bond with each other; and

B. Circle locks, on U and $\overline{U}$ quarks, cannot bond with each other.

Therefore, d and $\bar{d}$ quarks can only bond with U and $\overline{U}$ quarks, respectively, and vice versa.

The underlying structure and composition of the d and $\bar{d}$ quarks are predicated on both having two Z bosons, $(VV/PP)^{0}$; but differing from each other by the d having a W- boson, $(PP/PP)^{(-)1/3}$, present; while the $\bar{d}$ has a W+ boson, $(VV/VV)^{(+)1/3}$.

Similarly, the U and $\overline{U}$ quarks each possess only one Z boson, and differ from each other in that the U has two W+ bosons, while the $\overline{U}$ has two W- bosons.

Thus, these differences determine the respective arrow-key structure for the d and $\bar{d}$ quarks, while mandating the lock-like circles for the U and $\overline{U}$ quarks. Furthermore, the d and $\bar{d}$ quarks are the mirror-image of each other due to their charge difference, as are the U and $\overline{U}$ quarks for the same reason. The following two bonds, then, are precluded:

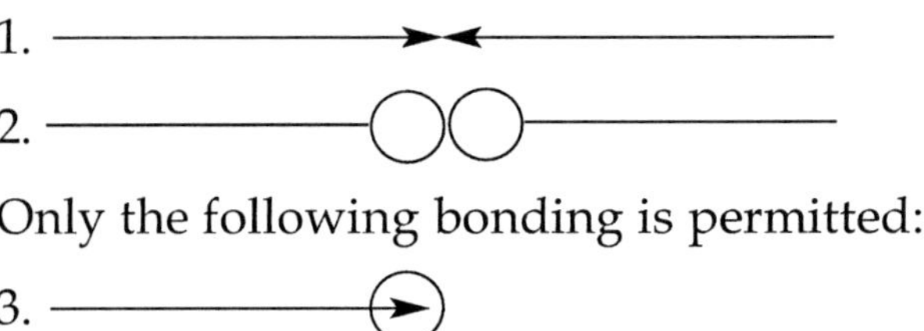

Only the following bonding is permitted:

3.

Lock and Key Structures

Of the many quark and g/ā meson interactions that occur in the E4 dimension, only four stable structures survive:

1) the proton shell/g/ā meson$^{(+)1}$,
2) the neutron shell/g/ā meson^{0},
3) the antiproton shell/g/ā meson$^{(-)1}$ and
4) the antineutron shell/g/ā meson^{0}.

(continued on p.176)

Diagram 8A

Proton shell/g/ā meson$^{(+)1}$, (UUd/g/ā)$^{(+)1}$

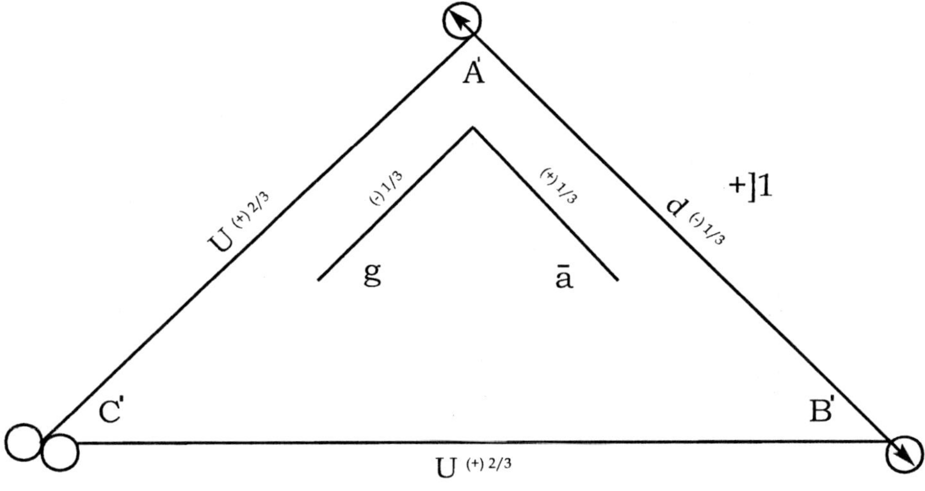

Diagram 8B

Neutron Shell/g/ā Meson^{0}, (Udd/g/ā)^{0}

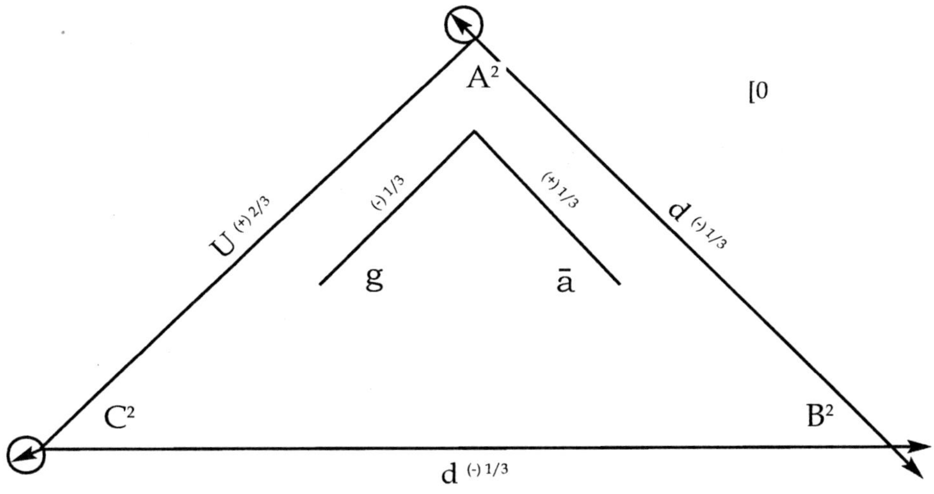

The first surviving stable structure (#1, above) is depicted in Diagram 8A, p.175, $(UUd/g/\bar{a})^{(+)1}$; while the second (#2, above) is depicted in Diagram 8B, p.175, $(Udd/g/\bar{a})^{0}$. The remaining two structures will not be illustrated, since both are antiparticles and both will soon disappear as a result of interactions with positrons, becoming primarily proton shell/g/ā mesons$^{(+)1}$.

Diagrams 8A and 8B both depict a $U^{(+)2/3}$ quark bonded to the $g^{(-)1/3}$ arm of the g/ā meson, while the $\bar{a}^{(+)1/3}$ arm has a $d^{(-)1/3}$ quark bonded to it; reflecting the opposites attract rule in E4.

The bottom quark, in each of the two diagramed structures shown, determines whether a proton shell/g/ā meson$^{(+)1}$ or a neutron shell/g/ā meson^{0} will result. Diagram 8A, having a $U^{(+)2/3}$ as the bottom quark, thereby is the former, $(UUd/g/\bar{a})^{(+)1}$; while the $d^{(-)1/3}$ quark in the same position causes the latter to be a $(Udd/g/\bar{a})^{0}$.

In Diagram 8A. p.175, two same type bonds are formed, (↘); each involving an arrow key, ↖ , penetrating a lock-circle, ○ .

Diagram 8C

Deuterium/g/ā Meson$^{(+)1}$

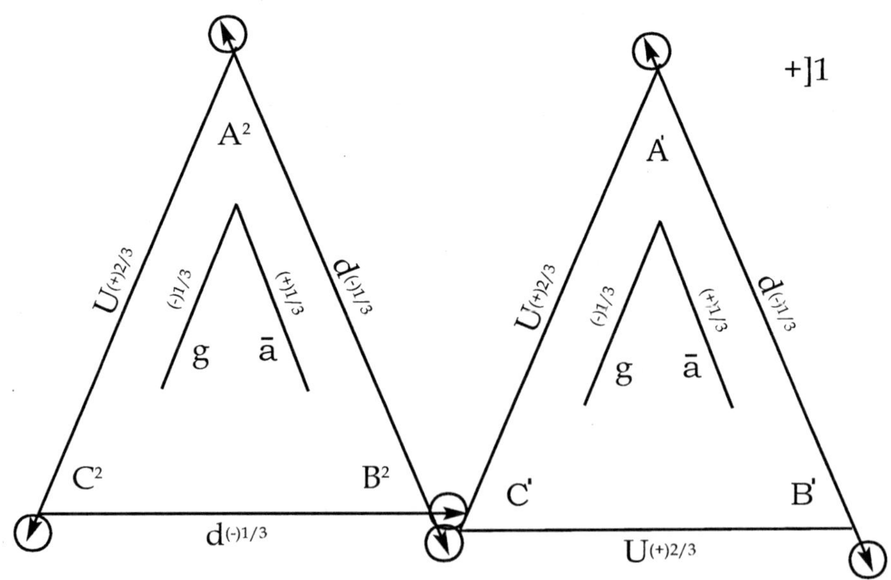

(Neutron Shell/g/ā Meson^{0}) (Proton Shell/g/ā Meson$^{(+)1}$)
$(Udd/g/\bar{a})^{0}$ $(UUd/g/\bar{a})^{(+)1}$

The single $d^{(-)1/3}$ quark present, on side A' B', has each of its two end arrow keys penetrate one lock-circle end of each of the two $U^{(+)2/3}$ quarks present (side C' A' and C' B'); thereby producing a bond at A' and B'.

Each of the two $U^{(+)2/3}$ quarks, therefore, has one unbonded lock-circle remaining, which are contiguous and are are found at C' : ○○.

The neutron shell/g/ā meson⁰, illustrated in Diagram 8B, also has two same type of bonds present, ⊗.

The single U(+)2/3 quark (side A^2 C^2) has each of it's two lock circle ends penetrated by a single arrow key from each of the two d(-)1/3 quarks present; producing a bond, ⊗, at A^2 and C^2.

An unbonded arrow key remaining from each of the two $d^{(-)1/3}$ quarks are both found at B^2, ↘→ in diagram 8B.

Since a lock-circle, ○ , cannot penetrate another lock-circle, a bond between two lock-circles is precluded. The arrow key, → , similarly cannot penetrate or bond with another arrow key, → .

If we, however, let the two unbonded lock-circles found at C', ○○, in Diagram 8A, $(UUd/g/ā)^{(+)1}$, contact the two unbonded arrow keys of the $(Udd/g/ā)^0$ found in diagram 8B, at B2, they will immediately bond together, producing two new ⊖ bonds, and a new and stable addition product, the deuterium/g/ā $meson^{(+)1}$, the second of three isotopes of the element hydrogen.

The structure of the deuterium /g/ā $meson^{(+)1}$ is illustrated in Diagram 8C. The penetration and bonding by the two arrow keys ↘→, situated at B^2 in the $UUd/g/ā^0$ into the two lock circles, ○○, present at C' in the $(UUd/g/ā)^{(+)1}$, has resulted in the structure depicted in Diagram 8C, p.176.

A variant of this type of bonding action can take place between a single proton shell/g/ā $meson^{(+)}$ and two neutron shell/g/ā $mesons^0$, instead of one neutron shell/g/ā meson.

This is depicted in Diagram 8D p.178, Tritium/g/ā $Meson^{(+)1}$. Here, a proton shell/g/ā $meson^{(+)1}$ has each of its two lock-like circles, ○○ , at C', penetrated and bonded by a single arrow key, → , from each of two neutron shell/g/ā $mesons^0$, at each of the same B^2 sites in each, instead of two arrow keys from a single neutron penetrating and bonding the two lock-like circles as in

Diagram 8D

Tritium/g/ā Meson$^{(+)1}$

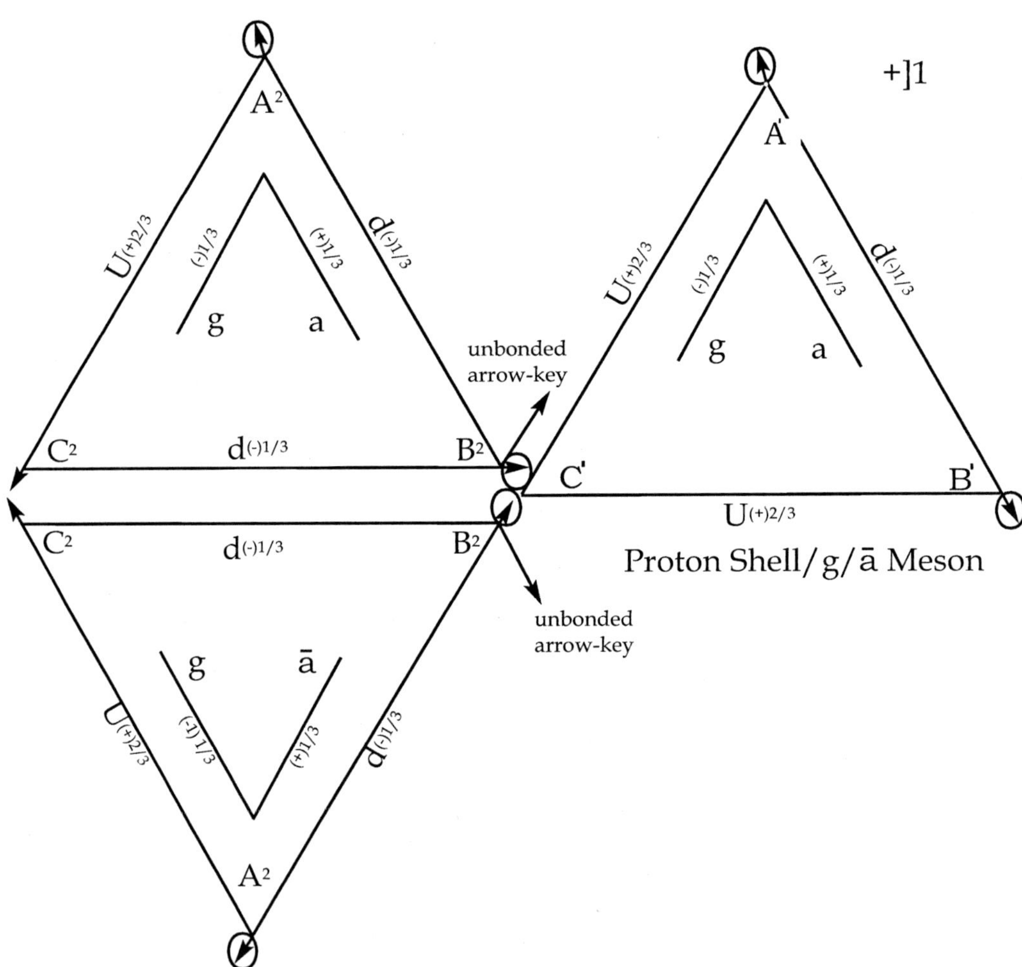

Diagram 8C. This also results in a single arrow key from each of the two neutron shell/g/ā mesons, (↗and↘) at positions B^2 in each, remaining unbonded at each B^2 position.

This is an unstable configuration, which leads to half of these tritium/g/ā meson$^{(+)1}$ units radioactively decaying in 12.5 years.

The cause of the radioactive decay can be an energy flux, which produces a twisting motion in one or both of the B^2 arrow keys embedded in their respective lock-circles of the proton shell/g/ā meson$^{(+)1}$, shown in diagram 8D.

If the energy flux is very great, both arrow keys can be pulled out of their lock-circles, producing a proton/g/ā meson$^{(+)1}$ and two neutron shell/g/ā mesons^0.

If the energy flux is less intense, only one of the arrow keys from a single neutron shell/g/ā meson^0 may be extracted from its lock-circle, resulting in a deuterium/g/ā meson$^{(+)1}$ and a single neutron shell/g/ā meson^0.

Z Boson Flexibility and Its Consequences

There are two types of g/ā mesons. The first type is formed in a magnetic dimension, above 10 million K, where the bond holding the gluon to the antigluon at its V-base is either a (+) to (+), or a (-) to (-) monopole-to-monopole bond, (chart VIII, S1-S4, p.36-37.)

The second type of g/ā meson is formed in an electric dimension at temperatures below 10 million K. Here, a (+) to (-) monopole-to-monopole bond is in place to hold the gluon and antigluon arms of the meson together. This is due to the monopoles in an electric dimension having to follow the opposites attract, like repels rule,(see chart VIII, U1-U5, p.38-40.)

This latter type bond is easily broken by an energy flux, causing the g/ā meson to be removed from the interior of a proton or neutron shell/g/ā meson. This is referred to as a proton or neutron decay process, respectively. Yet, the same (+) to (-) monopole-to-monopole bond between a d quark and a U quark in the proton shell or neutron shell, encircling the g/ā meson in the same type of electric dimension, remains unbroken by the same flux.

How can the same (+) to (-) bond holding a g/a meson together be so easily broken by a energy flux, and the same flux is not able to break the (+) to (-) type bond between a d quark and a U quark in a proton shell or neutron shell which surrounds the same g/ā meson?

The answer lies in the composition and structure of the gluon and antigluon versus the d quark and U quark.

The former has only W+ and W- bosons in its composition; while the latter is composed of W+ and W- bosons as well as at least one Z boson.

The Z boson (PP/VV) has the property of being flexible, while the W+ boson (VV/VV) and the W- boson (PP/PP) are inflexible.

In the face of an energy flux, structures containing Z bosons, such as d and U quarks, can flex, thus relieving a great stress on their structure, as well as the possible rupture in the bond structure at the point of maximum stress. Trees that bend or flex do not break.

The g/ā meson, on the other hand, containing only W+ bosons (VV/VV) and W- bosons (PP/PP), both of which have no flexing ability, can have a stressed bond in their structure break due to this same energy flux.

Another factor present which has an effect on these processes is the triangular structure of the proton shell or neutron shell, which gives great strength to these shell structures.

The g/ā meson with its two-part V-shaped structure lacks the great structural support inherent in the shell type triangle configuration.

Due to these structural differences, great energy fluxes, such as those produced by contacting positrons, produce quite complex results with many of the previously referenced structures. These will be examined separately.

Intact and Shell Protons and Neutrons

The proton shell/g/ā meson$^{(+)}$ will hereafter also be referred to as an "intact proton$^{(+)1}$"; and the neutron shell/g/ā meson^{0} will be called an "intact neutron^{0}".

Their counterparts which have had their g/ā mesons excised are respectively called a shell proton$^{(+)1}$ and a shell neutron^{0}, UUd and Udd respectively.

In the aftermath of certain reactions, an intact proton$^{(+)1}$ or an intact neutron^{0} may lose its central g/ā meson. The remaining three-part outer shell, either as a UUd or a Udd, respectively, continues on, seemingly unaffected by the excision of the g/ā meson from its formerly intact UUd/g/ā or Udd/g/ā structure.

Most of the reactions of the intact proton$^{(+)1}$ or neutron^{0} are carried out in the exact same way by the shell proton$^{(+)1}$ or neutron^{0}.

There is, however, one major exception. When intact protons$^{(+)1}$ and neturons^{0}, as well as shell protons$^{(+)1}$ and neutrons^{0}, are used to construct the various elements of the periodic table, the shell protons$^{(+)1}$ and neutrons^{0} therein are the only ones prone to a momentary breaking of one of their connecting double-quark bonds, which link and fuse together all of the protons and neutrons that form long chains in the element's structure. A break is generally due to one of the many types of energy fluxes, such as by the impingement of a positron$^{(+)1}$. The intact proton$^{(+)1}$ and intact neutron^{0} present in the same element's chain, however generally is completely unaffected by these same energy fluxes to which shell units are vulnerable.

If the force produced by an energy flux is projected onto a shell unit which is in a strategically vulnerable position in the structure of an element, a momentary break in the boson-to-boson bonds of a double-quark between, for example, two successive proton shells in the midst of a chain of intact protons and neutrons in an element will cause a catastrophic alteration in that element. The following is an example:

1) $${}^{232}_{92}\text{U} \longrightarrow {}^{208}_{82}\text{Pb} + {}^{24}_{10}\text{Na}$$

In the above example, two successive single shell protons, in the middle of a chain of predominately intact protons, which in successive order make up the spine of the uranium atom, ${}^{232}_{92}\text{U}$, has a nearby energy flux impact it.

This causes a great strain to be centered on the double-quark bond which fuses the two succesive shell protons together.

Due to the absence of a g/ā meson in both of these single shell protons, and the stabilizing reinforcement it had confered on both, the impact of the energy flux causes the fused double-quark structure to twist wildly as it absorbs the impact of the energy flux. As a result, the monopole components of the bosons in the double-quark are caused to momentarily break the bonds between them.

This action causes a momentary rupture and shutdown of the bond, as well as the component (+) 1 charge in both proton shells. When the violent twisting which caused the break ceases, the (+) 1 bond circuit and charge, in each of the two severed shell protons which had been turned off, are once more turned on in both proton shells.

The magnetic bond which was present between the above (+) 1 proton shell and its adjacent (+) 1 proton shell was forged in a magnetic dimension, the M11, where like attract and bond.

Now, however, the two newly recharged (+)1 proton shells are, in an electric dimension, the E11. The rules for bonding are: like repel, opposites attract.

Therefore, there is an immediate repulsion between these two newly restored (+) 1 charged proton shells, which were previously fused together as a double-quark, which also fused the chain together. This causes a split in the chain of protons and neutrons in the Uranium atom. Everything attached to one of the (+) 1 proton shells splits off and becomes a lead atom, ${}^{208}_{82}Pb$; while the other (+) 1 proton shell and everything attached to it become a new sodium atom, ${}^{24}_{10}Na$.

Another example of the effect of an energy flux on a shell proton is the following:

2) $${}^{226}_{88}Ra \rightarrow {}^{226}_{86}Rn + {}^{4}_{2}He^{(+)2}$$

Here, the radium atom (${}^{226}_{88}Ra$) has a central spine of mainly intact protons and neutrons; as well as side chains coming off the spine which have a large number of shell units composing them.

In example 3, p.183, a side chain coming off the central spine of intact protons and neutrons in ${}^{226}_{88}Ra$ is illustrated.

3) Radium ($^{226}_{88}$Ra)

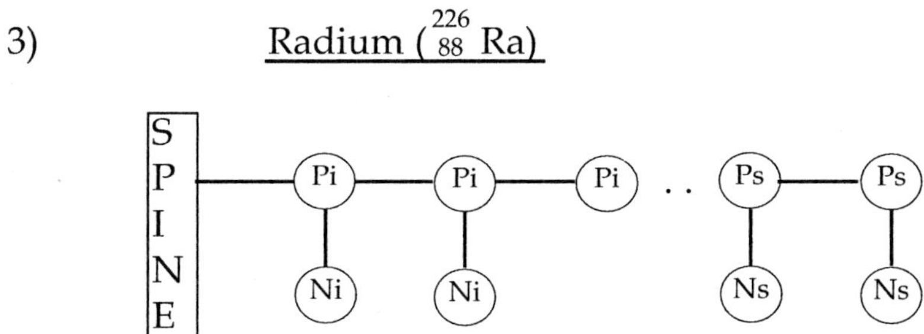

The intact protons and neutrons are labeled (Pi) and (Ni), respectively; the shell protons and neutrons are (Ps) and (Ns), respectively.

As in the previous uranium atom example, an energy flux causes an intense twisting motion therein, which momentarily severs the electric circuit in the double-quark splicing together the (Pi) . . (Ps), intact proton and shell proton, as illustrated in diagram 3 above; thus momentarily turning off the (+)1 charge in both protons. When the electric circuit and (+)1 charge in both protons return after the twisting motion ends, the two (+) 1 protons now must repel each other, due to like charge repulsion in an electric dimension, thereby breaking the (+) 1 to (+) 1 bond (. .) between the shell proton ((Ps)) and the intact proton ((Pi)) in the middle of the side chain. This causes the shell proton ((Ps)) and its attached end shell proton ((Ps)) and both of their attached bonded shell neutrons((Ns)) to separate from the ((Pi)) and the remainder of the side chain and move away.

This separated unit is an ionized helium nucleus, ($^{4}_{2}$He)$^{(+)2}$, or alpha particle. The remainder of the side chain, from which it split off, stays bonded to the spine of intact protons and neutrons, which thus relieved of two shell protons and two shell neutrons, is now a radon atom ($^{222}_{86}$Rn), instead of a radium atom ($^{226}_{88}$Ra).

There are numerous other examples of splitting of an element into two smaller elements.

All of these above reactions, however, are due to the impact of an energy flux upon a shell proton or shell neutron in the central spine or side chain of an element.

Strong Force

The sum total of all magnetic and electric forces and fields generated by respective V photons and P photons, comprising all of the structural photonic entities in a g/ā meson, in the form of magnetic and electric charges used in bonding itself together, as well as bonding to itself a three-quark shell entity: either a UUd (a shell proton), a Udd (a shell neutron), a $\overline{\text{UUd}}$ (a shell antiproton), or a $\overline{\text{Udd}}$ (a shell antineutron), is referenced as the strong force. Thus, the g/a mesons strong force stabilizes all of the double-quark bonds which fuse together the chains of intact protons and intact neutrons, preventing radioactive decay in an element.

Weak Force

The sum total of all magnetic and electric forces and fields generated by respective V photons and P photons, comprising all of the structural photonic entities in a unitary three-quark shell structure: either a UUd, a Udd, a $\overline{\text{UUd}}$, or a $\overline{\text{Udd}}$, in the form of magnetic and electric charges used in bonding together the three quarks into a shell unit, as well as bonding the shell unit to a g/ā meson, is referenced as the weak force. It can not prevent a break in a double-quark bond; thus, it can not prevent radioactive decay in an element.

Land of the Lost

The antiproton$^{(-)1}$ and antineutron^{0} structures are presented in Diagrams 9A, 9B, 9C and 9D (P.185.) 9B and 9D illustrate the intact proton$^{(+)1}$ and neutron^{0} structures using the arrow key and lock-circle charges in bond formation. However, all the antiprotons and antineurons will disappear from the existent universe's stage by way of reactions with positrons$^{(+)1}$, producing primarily protons in their place. Since their time in the existent universe is so limited, we have chosen not to further discuss them.

We will, for the same reason, also neglect many other entities that have similar short lifetimes: UUU, ddd, $\overline{\text{UUU}}$, ddd and other bits and pieces of photons, which also appear and then disappear as quickly, or are not a part of the structures of any extant existing elements (muons, taons, etc.).

Diagram 9

Lock and Key Structures

Antiproton$^{(-)1}$

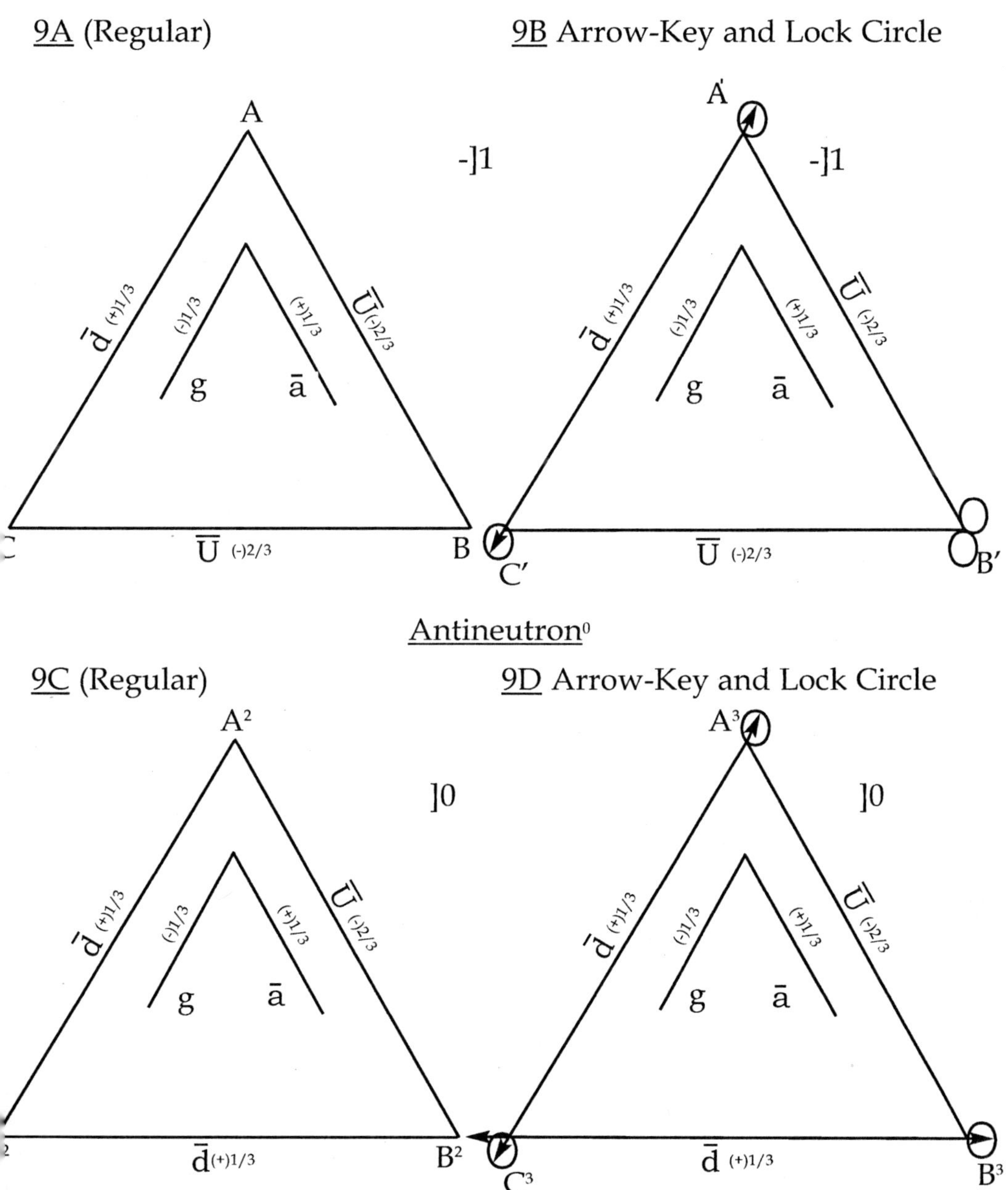

These rather fugitive entities, a zoo of very short-lived entites which never become a part of an existing element, are quite interesting. However, they must at this juncture be left to those specialists who are qualified to glean their vast riches.

Our purpose here is to deal only with those structures which constitute the actual structural matter of the existent Universe.

Hydrogen Isotope Construction

The proton$^{(+)1}$, the neutron^{0}, the antiproton$^{(-)1}$ and the antineutron^{0}, each with an intact g/ā meson, are produced in the E4 dimension. Then, each of these entities makes its exit out of the E4 as part of a separate and continuous stream, creating a new area, the E7. Here, each stream encounters and interacts, as follows:

I. The proton$^{(+)1}$ and antiproton$^{(-)1}$ streams intersect one another; individual units then bond together into neutral proton/antiproton^{0} units. These then flow towards a relatively cool outer area in the E7 (below 10 million K.) which abuts the M11, magnetic dimension, which contains polymerized magnetic electrons and polymerized magnetic positrons.

As these latter two polymerized units further cool, the polymerized magnetic electrons dip in temperatures to just below 10 million K., causing each of the six (-) 1/6 magnetic monopole charges, contained in each magnetic electron in the chain of such polymerized electrons, to now form an electric circuit between themselves and thereby exhibit a single electric (-) 1 charge.

It is this specific type of cooling action which causes the depolymerization of the chains of magnetic electrons into a stream of single, free (-) 1 , electric charged electrons to enter the electric E7 dimension.

This transformation occurs at the interface area between the magnetic M11 and the electric E7 dimensions; where the newly formed proton/antiproton^{0} entities are also present; resulting in the newly formed, free electrons$^{(-)1}$ contacting and then substituting themselves for the antiproton$^{(-)1}$ in the proton/antiproton^{0} units.

Thus, a new entity is created, the proton/electron^{0}, also called a hydrogen atom; while the antiproton$^{(-)1}$ is ejected into an adjoining area.

The ability of the electron$^{(-)1}$ to sever the bond between the proton/antiproton^{0} unit and then displace and substitute itself for the antiproton$^{(-)1}$, producing a bond with the proton$^{(+)1}$, is predicated on the size differential between the total surface areas of the electron$^{(-)1}$ and antiproton$^{(-)1}$.

The electron's (-) 1 charge is spread over its entire surface area, composed of twelve P photons$^{(-)1/12}$, which is relatively tiny as compared to the antiproton's$^{(-)1}$ huge surface area made up of thirty-six P photons$^{(-)1/12}$ and twenty-four V photons$^{(+)1/12}$.

The electron$^{(-)1}$ and the antiproton$^{(-)1}$, both projecting (-) 1 charges, are similarly attracted to the proton's (+) 1 charge in an E7 area. Both essentially compete for the proton's (+) 1 charge.

The electron will always win out due to its relative small size, which permits it to more easily and efficiently form a strong bond with the proton$^{(+)1}$. The huge size of the antiproton$^{(-)1}$ prevents an easy or efficient bonding. In fact, the bond formed with the proton$^{(+)1}$ was relatively quite weak,due to this factor.

Once the proton/electron unit, or the hydrogen atom $({}_{1}H^{1})^{0}$, is formed, it is a neutral or zero (0) charged unit, forcing it to stream out of the electric E7 dimension in order to find a new neutral-electric area for itself; the N1.

As it traverses the E7, the hydrogen atom attracts and then covalently bonds with another like hydrogen atom; thereby producing a more stable configuration, the neutral hydrogen molecule, or $({}_{1}H^{1})_{2}^{0}$, which will eventually accumulate in a new area, the N1 dimension.

II. The same dip in temperature in the M11 that permitted the depolymerization of the polymeric, magnetic electron chains into (-) 1 electrons, also causes a concomitant depolymerization of the parallel polymeric, positron chains present in the M11 into single (+) 1 positron units. Here, the six (+) 1/6 monopoles in each magnetic positron in the polymerized chain establishes an electric circuit between themselves, producing the (+) 1 charged positron.

Entering the E7, these are immediately attracted to the (-) 1 charged antiprotons, which have previously been severed and displaced from their proton$^{(+)1}$ units by the newly depolymerized electrons $^{(-)1}$.

As these two oppositely charged units contact one another, a violent reaction occurs in which both units are primarily converted into a proton$^{(+)1}$ and an electron$^{(-)1}$, which then instantly merge by bonding with each other into a neutral hydrogen atom^{0}.

These then follow the same path through the electric E7 as previously described, covalently bonding with another like neutral hydrogen atom, thereby producing a neutral hydrogen molecule, $({}_{1}H^{1})_{2}^{0}$, which also winds up in the N1 neutral-electric dimension.

III. Neutron shell/g/ā mesons^{0}, formed in the E4 dimension, also stream into the E7 dimension. Here, "hot" magnetic proton shell/g/ā mesons^{M}, above the 10 million K. threshold, which have just been created as a result of violent positron$^{(+)1}$ reactions with antiproton shell/g/ā mesons$^{(-)1}$, as well as violent positron reactions with antineutron shell/g/ā mesons^{0}, intercept and react react with the neutron shell/g/ā mesons^{0}, due to both being raised to temperatures above 10 million K., As a result, magnetic deuterium/g/ā mesons^{M} are produced, which, eventually, after cooling below 10 million K., form electric or ionic deuterium/g/ā mesons$^{(+)1}$, which are then neutralized by electrons$^{(-)1}$.

The resultant neutral deuterium/g/ā meson^{0} atoms then flow towards the N1 dimensional area. As they traverse the E7, each unit covalently bonds with another like unit, producing a neutral deuterium/g/ā meson^{0} molecule, $({}_{1}^{2}H)_{2}^{0}$, which eventually finds its way into the newly formed N1 dimension.

In a number of instances, however, two magnetic neutron shell/g/ā mesons^{M}, instead of one, bond with the magnetic proton shell/g/ā meson^{M}, creating a magnetic tritium shell/g/ā meson^{M}, which,when cooled into an ionic unit, is then similarly neutralized by an electron$^{(-)1}$ into a neutral tritium atom, $({}_{1}^{3}H)_{1}^{0}$.

It then also flows toward the N1, and is covalently bonded to another like unit, producing a tritium shell/g/ā meson^{0} molecule, $({}_{1}^{3}H)_{2}^{0}$. It then mixes into to the other two hydrogen isotope molecules in the N1: $({}^{1}_{1}H)^{0}_{2}$ and $({}^{2}_{1}H)^{0}_{2}$.

$({}^{3}_{1}H)^{0}_{2}$ is unstable, with a half-life of 12.5 years, which can decay into either a $({}^{2}_{1}H)^{0}_{2}$ or a $({}^{1}_{1}H)^{0}_{2}$ molecule.

IV. The antiproton shell/g/ā meson$^{(-)1}$, also called an antiproton/g/ā meson$^{(-)1}$, and the antineutron shell/g/ā meson, also referred to as an antineutron/g/ā meson^{0}, are also formed in the E4, and stream into the E7. Here, positrons (e+) react with them, primarily producing proton/g/ā mesons$^{(+)1}$ in both cases; as well as electrons in the former reaction (Charts XV.1 to 5, p.92-96.) and antineutrinos ($\overline{V}$) in the latter (Charts XVI.1 to 5, p.99-102.)

In a number of cases, however, the former reaction produces the following: $2e^{+}+2e^{-}+V+\overline{V}$ (Charts XV.6A and 6B,p.98), instead of the proton/g/ā$^{(+)1}$ and an electron (e-); while the latter reaction produces the following: $2e^{+}+e^{-}+V+2\overline{V}$ (Charts XVII.1 to 3, p.104-106), instead of the proton/g/ā$^{(+)1}$ and the antineutrino $\overline{V}$.

The proton/g/ā$^{(+)1}$ units generated in the positron (e+) reactions, cited above, emerge from these two reactions as mini-M11 magnetic "hot protons", both above 10 million K., due to the intensity of the above reactions. No other reaction is as exothermic in nature. These "hot protons" emerge from the reaction as magnetic protons/g/ā mesons^{M}, transferring heat energy to any neutron/g/ā meson^{0}, or two neutron/g/ā mesons^{0}, they may encounter just after emerging from the reaction; causing the neutron/g/ā mesons^{0} to heat up to over 10 million K., thus becoming mini-M11 magnetic "hot neutron" units. These are then able to fuse with the magnetic proton/g/ā meson^{M} and form either a magnetic deuterium/g/ā meson^{M} or a magnetic tritium/g/ā meson^{M}, both of which when cooled to under 10 million K. become ions in the E7. These are then neutralized by electrons extant, and transformed into atoms.

The three hydrogen atom isotopes from the above three reactions will covalently bond with like atoms on their way to the N1, creating new molecules of hydrogen, deuterium and tritium.

In summary, the protons$^{(+)1}$ generated in the reaction between the positron$^{(+)1}$ and the antiproton$^{(-)1}$ and between the positron$^{(+)1}$ and the antineutron^{0} are "hot" protons$^{(+)1}$, whose temperature exceeds 10 million K. This is due to the violent exothermic energy reactions in both cases; which, taken together with the already near 10 million K. temperature extant in the area of their interaction, causes the area surrounding each newly generated proton$^{(+)1}$ to exceed 10 million K. This converts the electric E7 proton$^{(+)1}$ into an M11 magnetic proton^{M}; which then allows for an easy transformation into magnetic deuterium and tritium, when it intercepts a nearby neutron or neutrons, respectively.

Since all four entities which exited the E4, flowing into the E7 (protons$^{(+)1}$, neutrons^{0}, antiprotons$^{(-)1}$ and antineutrons^{0}), were produced in equal quantities, it thus follows that nearly half of the protons$^{(+)1}$ which eventually appear in the E7 were produced through catastrophic interactions between positrons$^{(+)1}$ and the two antiparticles. These "hot" protons' immediate interactions with extant neutrons in the E7, thereby, make it possible for nearly all of the deuterium^{0} and tritium^{0} molecules extant to be derived from the protons produced in these catastrophic reactions.

These three isotopes of hydrogen are the essential raw materials which will be converted into all of the elements in the periodic table.

We will next examine how this occurred in the course of the creation of our present solar system.

Recapitulation of E7 Reactions

Preliminary to a discussion of the element building mechanism, a recapitulation of many types of reactions (12) which occur in the E7 dimension should be useful.

Diagram 10, p.191, which follows, lists two types of reactions:

A. Primary: 1, 3 , 7

B. Secondary: 2, 4, 5, 6, 8, 9, 10, 11, 12

Primary reactions 1 and 3 show the conversion of antiprotons$^{(-)1}$ and antineutrons^{0} by positrons$^{(+)1}$ into proton/g/ā$^{(+)1}$ units; while 7 depicts the neutralization of the proton/g/ā$^{(+)1}$ by an electron$^{(-)1}$ into a hydrogen atom^0. If, however, a deuterium/g/ā$^{(+)1}$, or tritium/g/ā$^{(+)1}$, is substituted for the proton/g/ā$^{(+)1}$, then the reaction yields a respective deuterium atom^0 or tritium atom^0.

Secondary reactions illustrate how complex the E7 is. Some of these reactions, such as 2 and 4, have been extensively looked at as alternative routes following respective antiproton$^{(-)1}$ and antineutron^0 reactions.

Other secondary reactions are presented to indicate the vast number of possible interactions that may occur, which ensure that the existent universe is not a homogenous place.

When all of the positrons$^{(+)1}$ (e+), antiprotons$^{(-)1}$, antineutrons^0, neutrinos (V_E) and antineutrinos ($\overline{V}_E$) have been used up in the reactions shown in Diagram 10, only the three molecular isotopes of hydrogen remain in the N1 dimension of the existent universe.

We will now discuss the very elegant mechanism which transforms the three isotopes of the element hydrogen into all of the elements of the periodic table.

Diagram 10

Reactions in E7

Positron$^{(+)1}$ (e^+)
1. e^+ + antiproton/g/ā$^{(-)1}$ → proton/g/ā$^{(+)1}$ + e^-
2. e^+ + antiproton/g/ā$^{(-)1}$ → $2e^+ + 2e^- + V + \overline{V}$
3. e^+ + antineutron/g/ā^0 → proton/g/ā$^{(+)1}$ + $\overline{V}$
4. e^+ + antineutron/g/ā^0 → $2e^+ + e^- + V + 2\,\overline{V}$
5. e^+ + neutron/g/ā^0 → proton/g/ā$^{(+)1}$ + $\overline{V}$

Proton/g/ā$^{(+)1}$ (UUd/g/ā)$^{(+)1}$
6. proton/g/ā$^{(+)1}$ + e^- → neutron/g/ā + V
7. proton/g/ā$^{(+)1}$ + e^- → hydrogen atom^0 ($_1H^1$)^0
8. proton/g/ā$^{(+)1}$ + $\overline{V}$ → neutron/g/ā^0 + e^+

Neutron/g/ā^0 (Udd/g/ā)^0
9. neutron/g/ā^0 + V → proton/g/ā$^{(+)1}$ + e^-

10. neutron/g/$\bar{a}^0$ + e^+ → proton/g/$\bar{a}^{(+)1}$ + $\overline{V}$
11. neutron/g/$\bar{a}^0$ → proton/g/$\bar{a}^{(+)1}$ + e^-

Electron(-)1 (e-)
12. e- + proton/g/$\bar{a}^{(+)1}$ → neutron/g/$\bar{a}^0$ + V

The Mechanism of Element Synthesis

The key step in the process of synthesizing all of the isotopes of all of the elements found in the periodic table, excluding the hydrogen isotope of hydrogen and the neutron, which is treated herein as an isotope, is the complete turning off of all of the electric circuits which produce the electric charges and bonds in all isotopes of all of the lighter elements that are to be spliced together in order to create new and heavier elements.

This inactivation of all the electric circuits only occurs at temperatures greater than approximately 10 million K. Those circuits rendered inactive due to such temperatures are:

I. Electric circuits present between paired types of monopoles: 1) between each of two monopoles, each having a (+) 1/6 charge, producing an an electric W+ boson, $(VV/VV)^{(+)1/3}$: 2) between each of two monopoles having a (-) 1/6 charge, producing electric W- bosons, $(PP/PP)^{(-)1/3}$; and 3) between one monopole having a (+) 1/6 charge, the other a (-) 1/6 charge, creating an electric Z boson, $(VV/PP)^0$.

II. Electric circuits between a triad of bosons, which produce quarks, gluons and antigluons:

1. Up quarks: two W+ bosons(VV/VV) and one Z boson (VV/PP), $[(VV/VV)(VV/VV)(VV/PP)]^{(+)2/3}$;
2. Down quarks: one W- boson (PP/PP) and two Z bosons, $[(PP/PP)(VV/PP)(VV/PP)]^{(-)1/3}$;
3. Gluons: two W- bosons and one W+ boson, $[(PP/PP)(PP/PP)(VV/VV)]^{(-)1/3}$;
4. Antigluons: two W+ bosons and one W- boson, $[(VV/VV)(VV/VV)(PP/PP)]^{(+)1/3}$.

III. Electric circuits between a single gluon and a single antigluon, producing a V-shaped gluon/antigluon meson.

IV. Electric circuits between a trio of quarks, creating a proton shell, or a neutron shell, $(UUd)^{(+)1}$ or $(Udd)^{0}$, respectively:

1. A proton shell: two up quarks (U) and one down quark (d), $UUd^{(+)1}$;
2. A neutron shell: one up quark (U) and two down quarks (d), Udd^{0}.

The above two entities can exist as independent shell units, $UUd^{(+)1}$ or Udd^{0}, or each can establish circuits and bonds with a g/ā meson^{0} at their center, creating a UUd/g/ā$^{(+)1}$ or Udd/g/ā^{0} entity.

When all of the foregoing electric circuits are turned off due to temperatures which exceed approximately 10 million K., fusion of elements can proceed.

What remains in structural entities after electric charge inactivation has occurred are exclusively magnetic charged monopoles: (+) types (VV) and (-) types (PP), exhibiting (+) 1/6 and (-) 1/6 charges, respectively.

Additionally, the rules of attraction and bonding are now reversed, as like attracts, opposites repel, as the dimension changes from electric to magnetic, E7 to M11, due to 10 milllion K.

All of the structures previously held together in the E7 by electric circuits and bonds now are kept in place by a magnetic attraction between monopoles, which cannot form electric circuits between themselves in the now magnetic M11. Thus, all magnetic monopoles in a magnetic dimension are not additive in their charges. Additive charges can only take place in electric dimensions.

The proton/g/ā$^{(+)1}$, depicted in Diagram 11-E.1, p.194, for example, illustrates how the electric type structure is constructed.

It is composed of three quarks: two up types, each at a (+) 2/3 charge: a UR1r and a UR2; and one down type, a dY1, with a (-) 1/3 charge. The entire structure is designated as a UUd/g/ā$^{(+)1}$ and has a net additive charge of (+) 1 [(+)2/3) + (+)2/3) + (-)1/3)].

Although we show the presence of a gluon, (g), and an antigluon (ā), as the two parts of a g/ā meson in the above

UUd/g/$\bar{a}^{(+)1}$, they do not partake in any of the reactions to be cited in this section; and, therefore, their structure will not require further discussion in the reaction context.

The neutron g/$\bar{a}^{0}$ will, however, partake in our reactions and is presented in Diagram 11-E.2, p194. Here, it is shown to be assembled from three quarks: two down types, each with a (-) 1/3 charge: a dY1 and a dB1; and one up type at (+) 2/3 charge: a UR1r. The entire structure is designated as a Udd/g/$\bar{a}^{0}$, with an additive

Diagram 11-E (Electric 1 and 2)

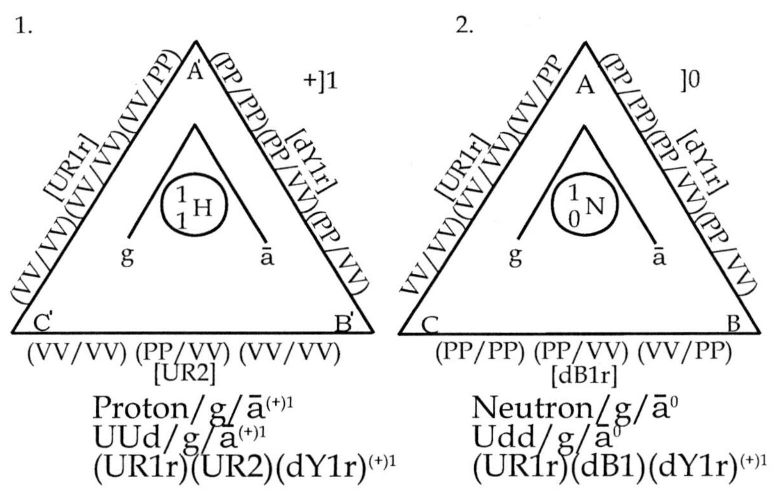

Diagram 11-M (Magnetic 1 and 2)

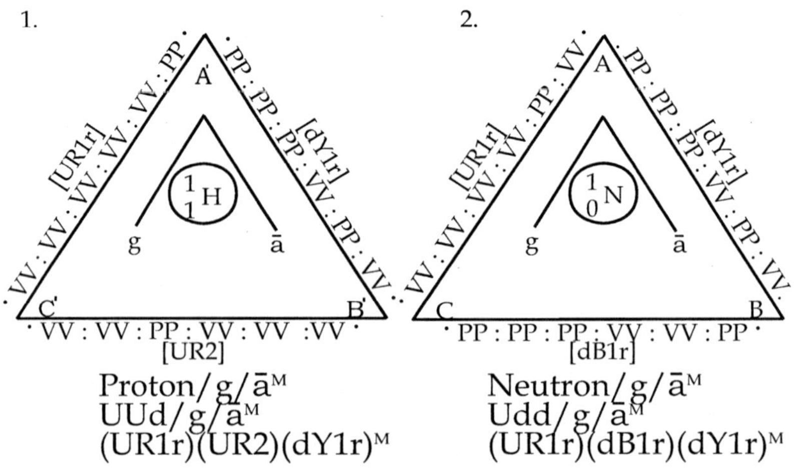

charge of 0 [(-)1/3) + (-)1/3) + (+)2/3)]. The g/ā meson herein also does not partake in any of the reactions.

Each up quark, $[(VV/VV)(VV/VV)(PP/VV)]^{(+)2/3}$, contains two W+ bosons, $(VV/VV)^{(+)1/3}$, in addition to a single Z boson, $(PP/VV)^{0}$, an additive (+) 2/3 [(+) 1/6 (+) 1/6 (+) 1/6 (+) 1/6 (-) 1/6 (+) 1/6)].

The down quark, $[(PP/PP)(PP/VV)(PP/VV)]^{(-)1/3}$, has a single W-boson, $(PP/PP)^{(-)1/3}$, in addition to two Z bosons, an additive (-) 1/3 [(-) 1/6 (-) 1/6 (-) 1/6 (+) 1/6 (-) 1/6 + 1/6)].

Diagram 11-E.3

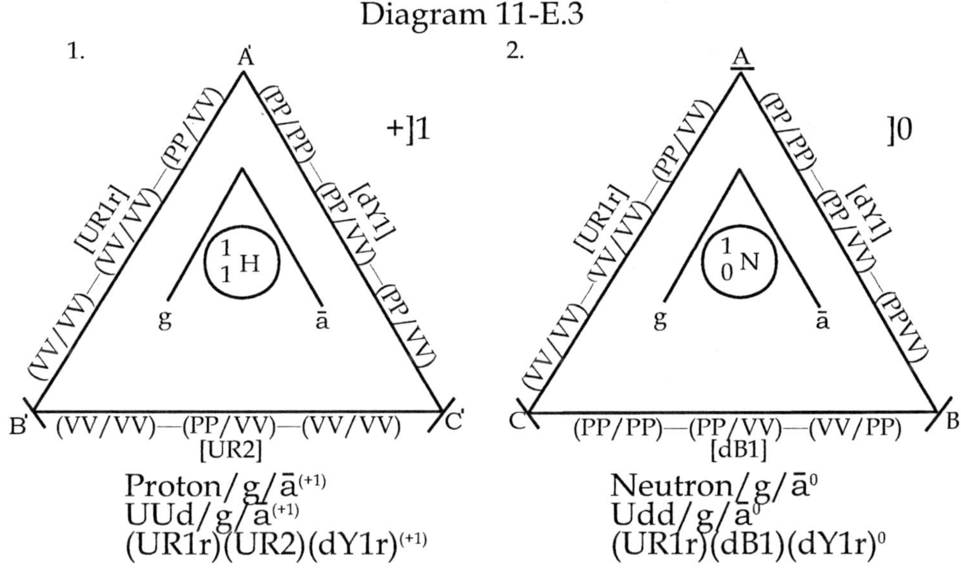

Proton/g/ā$^{(+1)}$
UUd/g/ā$^{(+1)}$
(UR1r)(UR2)(dY1r)$^{(+1)}$

Neutron/g/ā^{0}
Udd/g/ā^{0}
(UR1r)(dB1)(dY1r)^{0}

The electric type bond between any two monopoles in all of the above electric units is indicated by the / symbol.

Thus, between two (+) monopoles, $(VV)^{(+)1/6}$, an / (electric bond) produces a W+ boson, $(VV/VV)^{(+)1/3}$; between two (-) monopoles, $(PP)^{(-)1/6}$, an / (electric bond) signifies the creation of a W- boson, $(PP/PP)^{(-)1/3}$; and an / (electric bond) between a (-) monopole, $(PP)^{(-)1/6}$, and a (+) monopole, $(VV)^{(+)1/6}$, a Z boson, $(PP/VV)^{0}$, emerges.

There are also present nine electric boson-to-boson bonds, depicted by the symbol) (in each proton shell$^{(+)1}$ or neutron shell^{0}. Each of these are connected by a line,)—(, in Diagram 11-E.3, above.

There are two)—(in each of the three quarks, for a total of six, as well as three more)—(, each between the two ends of successive quarks; (A, Á, B́, etc) a grand total of nine in each of the proton and neutron shells

When all of the foregoing categories of electric circuits, charges and bonds have finally been turned off, due to reaching in excess of 10 million K., fusion of elements can then take place in

Diagram 11-M-A

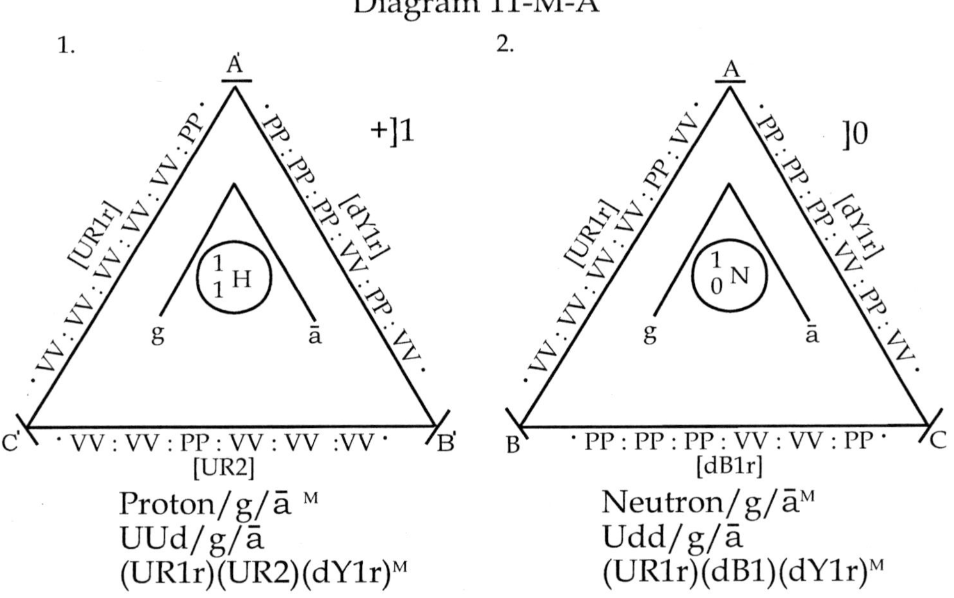

Proton/g/ā [M]
UUd/g/ā
(UR1r)(UR2)(dY1r)[M]

Neutron/g/ā[M]
Udd/g/ā
(UR1r)(dB1)(dY1r)[M]

what has become a magnetic phase dimension.

This is illustrated by comparing Diagrams 11-E-D, p.202, and 11- M-D, p.201,. Here we note that the electric bond symbols, / and) (, in Diagram 11-E-D, have been replaced by a two dot magnetic symbol, : , for the former (/); and a single dot, . , for each of the two latter symbols,) and (, in Diagram 11-M-D.

Also in 11-M-D, p.202, within each magnetic quark, each set of two electric bond symbols,) and (, is replaced by a single magnetic dot for each of the two, or a double dot : to represent both; while the very last) or (is rendered by a single dot for each. However, in Diagram 11-M-A, above, these single dots between quarks are connected by a line (—) above each letter (A, A[1], B[1], etc) to indicate that they are in reality magnetically bonded together to form a double dot, : ; three of which hold together each proton/g/$\bar{a}^{M}$ and each neutron/g/$\bar{a}^{M}$.

For example, the electric UR1r, up quark, seen in Diagrams 11-E.1 and 11-E.2, p.194, is depicted as follows:

1. (VV/VV)(VV/VV)(PP/VV)

The magnetic UR1r equivalent is depicted in the magnetic diagrams, 11-M.1 and 11-M.2, p.194, as:

2. · VV : VV : VV : VV : PP : VV ·

Each (-) magnetic monopole, · PP ·, and (+) magnetic monopole, · VV ·, shown above, has two magnetic centers of attraction, each designated by a single dot, ·, on each of its two sides, as exhibited below:

3. · PP · and · VV ·

When an electric type Z boson, (PP/VV), is transformed into its magnetic type equivalent, the following structure is produced:

4. · PP : VV ·

The middle two dots, : , above, represent a merger of one dot form the · PP · and one from the · VV ·, producing a magnetic bond.

The magnetic Z boson, shown in 4. above, still projects a single magnetic dot, indicative of its attraction and bonding potential, on either of its two sides. Each dot can further bond with another magnetic dot projected from any of the three magnetic boson types: W+ boson (· VV : VV ·), W- boson (· PP : PP ·) or Z boson (· PP : VV ·). The end product of such a union of three magnetic bosons is a magnetic quark.

The monopole charge components of a magnetic quark are never additive in a magnetic dimension (M11). It is only when electric circuits can be formed between monopole charges in an electric dimension (E7) that the monopoles' (+) 1/6 or (-) 1/6 charges become additive, so that an electric up quark is (+) 2/3 and an electric down quark is (-) 1/3.

All electric monopole-to-monopole bonds, / and)(, produce electric bosons within a quark component of each proton/g/$\bar{a}^{(+)1}$ and neutron/g/$\bar{a}^{0}$, Diagrams 11-E.1 and 11-E.2, p.195, respectively. The magnetic monopole-to-monopole bonds, : , which produce magnetic bosons within the quark components of a magnetic proton/g/$\bar{a}^M$ (Diagram 11-M.1),p.194 and within a magnetic neutron/g/$\bar{a}^M$ (Diagram 11-M.2), p.194 are classified as "Intraquark" bonds and bosons. These "intraquark" bonds and bosons only appear "within" each quark component of each proton or neutron shell; but they can never appear "between" the monopole components of a boson in a double-quark component of a proton shell and the same type of monopole components of the same type of boson in a neutron shell, or another proton shell.

Only vertical "interquark" bonds, creating vertical "interquark" bosons between two same types of quarks, thus producing a double-quark structure, can splice together a proton shell and a neutron shell; or a proton shell and a proton shell; or a neutron shell and a neutron shell. The procedure for these three types of synthesis, producing a double-quark will be examined in the following section, "Element Synthesis Procedure".

It should be first noted, however, that "intraquark" bosons can be either a W+, W- and Z type.

"Interquark" bosons can only be a W+ or W- type, and they can only be formed between two same quark types; each same quark being a constituent of a different shell, eg., a proton and a neutron; thereby creating a new type of quark between the two same quarks, the double-quark. This newly constructed double-quark contains six vertically bonded "interquark" bosons, made by vertically cross bonding the three "intraquark" bosons in a quark from the proton shell with the three "intraquark" bosons from the same type of quark in the neutron shell.

The double-quark is then the splice that creates, for example, the deuterium unit out of a free proton and a free neutron; or the same double-quark can link two smaller and lighter isotopes of elements together to produce a larger and heavier atom, such as an iron atom ($^{56}_{26}Fe$) or a uranium atom ($^{238}_{92}U$).

Element Synthesis Procedure

We will first illustrate this procedure by using a proton/g/$\bar{a}^{(+)1}$ and a neutron/g/$\bar{a}^0$ in a fusion reaction which produces deuterium/g/$\bar{a}^{(+)1}$, the second stable isotope of the element hydrogen. Although the neutron/g/ā^0 is technically not an isotope of an element, it will, however, be treated herein as if it were. The former is depicted in the previous Diagram 11-E.1, p.194, while the latter is depicted in Diagram 11-E.2. Each of these two entities is an electric charged structure, each containing an identical quark constituent, the UR1r, up quark. Both entities exist in the E7.

At a temperature in excess of 10 million K., the two electric charged entities are transformed into magnetic charged structures (M11), as respectively illustrated in previous Diagrams 11-M.1 and 12-M.2, (p.194.)

All of the bonds and charges in the quarks depicted in these latter two diagrams are now magnetic (:), thus permitting each of their UR1r, up quarks, in each proton and each neutron shell, to become contiguous, as their normally repelling electric charges have now completely disappeared. This contact would occur in a collision between both.

Also, the plus 10 million K. causes the conversion of the electric E7 area into a magnetic M11 area; as well as reversing the rules of attraction and bonding to : like attract, opposites repel.

As a direct result, the magnetic Z boson, · PP : VV ·, in each of these two now contiguous UR1r, up quark constituents of the proton/g/ā^M and the neutron/g/ā^M, due to the opposites repel rule in effect in this now magnetic dimension (M11), has each of its PP and VV monopoles repel each other (see Diagram 11-M-D), p.201.

This causes the distance to increase between the center of the PP monopole and the center of its two side VV monopoles in each of the two UR1r. Due to this, all of the monopole components of the UR1r in the proton/g/ā^M and the UR1r in the neutron/g/ā^M become unstable.

Both of these now contiguous and unstable horizontal positioned UR1r quarks immediately vertically cross-link and bond each of their PP monopoles, producing a stable vertical · PP : PP ·, W- boson between them. This is then followed by a cascading of the remaining horizontal VV monopoles in both to vertically cross-link and bond, creating five new and stable vertical · VV : VV ·, W+ bosons, between the two former horizontal quarks. This structure is now called a double-quark unit (DQ).

These six new vertical "interquark" bosons now splice together and thereby fuse the two former horizontal and separate UR1r quarks and each of their three "intraquark" bosons, as well as their host proton/g/a^{M} and neutron/g/a^{M}, into a single, new magnetic isotope of hydrogen, deuterium/g/a^{M} (Diagram 11-M-D), p.201.

This transformation can be summed up in terms of boson changes. In the proton/g/$\bar{a}^{M}$ and neutron/g/$\bar{a}^{M}$, each of their horizontal UR1r, up quarks, is composed of three "intraquark" bosons: two W+ and one Z. When these two horizontal UR1r, up quarks impact each other, six vertical "interquark" bosons are formed between each of their three "intraquark" horizontal bosons, creating a new combined and fused structure, the "double-quark".

In their magnetic form, all double-quark bonds are stable. It is only when they and their hosts are transformed into electric ions, neutral atoms or molecules, that some of them become unstable, with a specific half-life, and are considered to be radioactive. Deuterium ions, containing one neutron, are always stable; while tritium ions, containing two neutrons, are not, having a half-life of approximately 12.5 years. Yet, if magnetic tritium^{M} is reacted with another magnetic entity, such as magnetic helium 4^{M}, in a magnetic dimension, the magnetic end product, lithium 7^{M}, and its ionic and atomic forms are all completely stable.

However, if a tritium^{M} unit and a deuterium^{M} unit are coupled together, their end product, helium 5^{M}, in its ionic and atomic forms, is radioactive and disintegrates in a specific half-life.

It is the stresses within any end product which determines its stability or instability. Laboratory observations have determined the half-lives of almost all isotopes; or their absence; the latter then indicates a stable,non-radioactive isotope.

The magnetic deuterium^{M} isotope can be seen in Diagram 11-M-D (p.201). Note the six vertical magnetic "interquark" bosons between the UR1r, side AB, and the other UR1r, side A'B': one magnetic W- boson (PP : PP) and five magnetic W+ bosons (VV : VV). This is a magnetic double-quark which fuses the M-proton and M-neutron into a magnetic deuterium^{M} isotope $(^{2}_{1}H)^{M}$. All "double-quarks are labled "DQ" herein.

When the extant temperature in the M11 falls below 10 million

K., it transforms the M11 into the E7, whereupon each of the <u>magnetic</u> six vertical "interquark" boson bonds (· ·), in Diagram 11-M-D, are instantly transformed into six <u>electric</u> "interquark" boson bonds, with each : replaced by a —— symbol, showing that an electric bond is in place. This can be seen in Diagram 11-E-D, p.202, between sides AB and A'B'. In all of the "intraquark" bosons bonds, the electric bond symbols, (/) and)(denote the same type of bond as the ——, but <u>differentiate</u> visually between "<u>intraquark</u>" and "<u>interquark</u>" bonds. Additionally, in Diagram 11-M-D, below, the two unreacted quarks in the M- proton: the dY1r and UR2, as well as the two unreacted quarks in the M- neutron: the dY1r and the dB1r, each still contain their original three "intraquark" bosons, each bonded to another boson by magnetic "intraquark" bonds (:).

When the electric circuits are produced between monopoles, as the M11 is transformed into E7, from magnetic to electric space, it causes the isolated magnetic monopole charges of (+) 1/6 and

Diagram 11-M-D
M-Deuterium/g/ā ; ($^{2}_{1}$H)^{M}

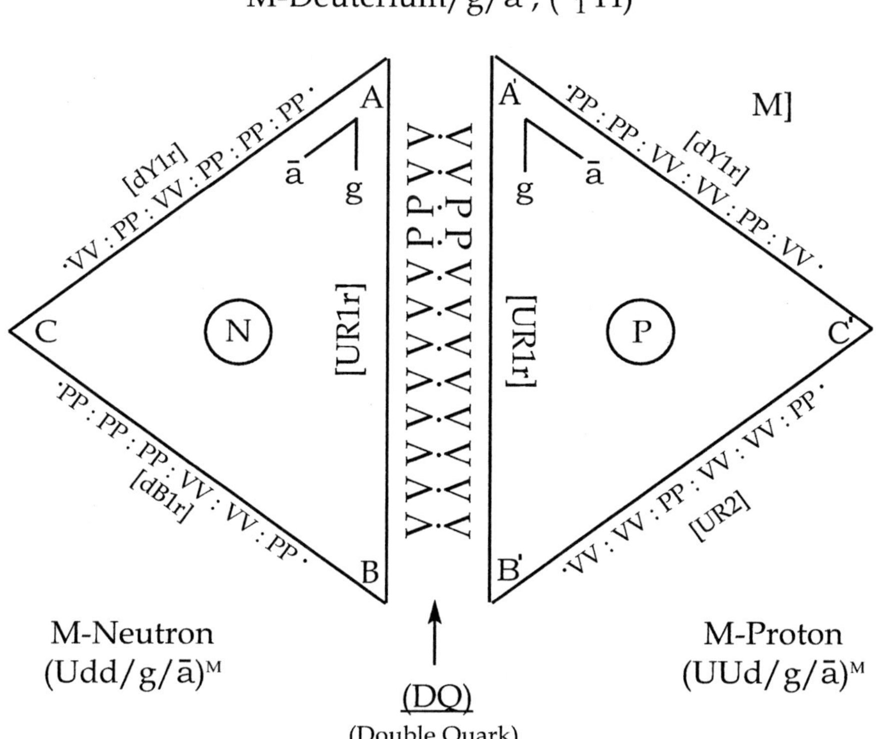

Diagram 11-E-D
Deuterium/g/ā$^{(+1)}$; ($^{2}_{1}$H)$^{(+1)}$

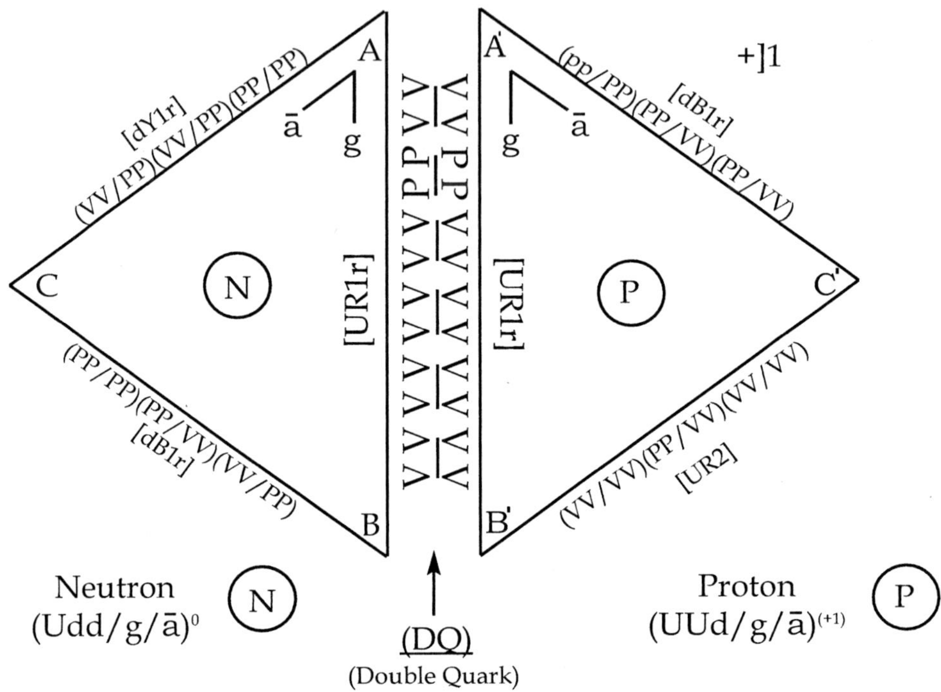

(-) 1/6 to become additive. Thus, the two individual VV$^{(+)1/6}$ are wired together to produce a (+) 1/3 W+ boson (VV/VV); while the two PP$^{(-)1/6}$ individual monopole units are similarly connected to form a W- boson (PP/PP) with a (-) 1/3 charge. Finally, the Z boson's PP$^{(+)1/6}$ and VV$^{(-)1/6}$ neutralize each other's charge, resulting in a zero (0) charged electric unit, (PP/VV)^{0}.

As the above magnetic monopoles are wired together together, (11-M-D), to create electric bosons, (11-E-D), their actual bond structures are not changed, as was the case when the electric monopoles lost their circuit connections when the temperature exceeded 10 million K. and they became magnetic type units. The magnetic bonds are not changed in any way in the transformation from magnetic to electric space; which also changes the bonding rules from like attracts, opposites repel to opposites attract, like repels.

It is only when electric circuits between electric monopoles disappear in the change from E7 to M11 that the newly formed magnetic monopoles can reverse how they repel or attract one another; as well as being able to alter the distances between themselves; as was the case in the PP monopole and its two side VV monopoles repelling each other as explained earlier. This latter type of change cannot take place in the transition from M11 to E7, as only electric circuits are put in place and magnetic bonds formed in M11 become electric, precluding any structural change. The only change that takes place when the M11 transforms into the E7 is that electric charges become additive, producing (+) or (-) 1/3, or higher charges.

Tritium/g/ā^{M} ($^{3}_{1}$H)^{M}

The last of the three isotopes of hydrogen, tritium/g/ā$^{(+)1}$, ($^{3}_{1}$H)$^{(+)1}$, is first synthesized in the magnetic M11 dimension.

Temperature is the critical factor in element synthesis. For example, the conversion rate in terms of hydrogen to helium at approximately 10-15 million K. is slow. The synthesis of elements beyond helium at this temperature range, however, is extremely slow, if not rare.

It is only when we reach the 1 billion K. range, or higher, as the result of a special event, a supernova or the like, that the rate of reaction of the elements is accelerated to such an extraordinary degree that great quantities of elements are produced, especially of the more complex and heavier types. We shall, however, discuss this in greater detail later.

Diagram12-III-D (p.205) depicts the final magnetic structure of the tritium/g/ā^{M} isotope of hydrogen after its formation as a result of two magnetic neutron/g/ā^{M} units (M-neutron, Ⓜ) having slammed into a single magnetic proton unit (M-proton, Ⓟ).

The M-neutron, at the left of the central M-proton, is the exact same type as was previously depicted as slamming into an M-proton in our description of the creation of M-deuterium/g/ā^{M}, (11-M-D, p.201).

It attaches itself to the M-proton, in the exact same process as described previously, by vertically cross-bonding each of six monopoles, contained in the three horizontal "intraquark"bosons, in each of two contiguous and opposite-facing up quarks (UR1r), resulting in the creationof six new vertical"interquark" bosons, the component structures of a new type of quark, the double-quark (DQ). It splices together the M-proton /g/ā/^{M} and the M-neutron/g/ā^{M} into a stable and non-seperable magnetic isotope of hydrogen, M-deuterium /g/ā.

On the right side of the M-proton, just described above, another M-neutron/g/ā^{M} of the exact same type as slammed into the M-proton/g/ā^{M} on the left side, also concurrently slams into it. However, it is the dY1r, down quark, on side A'C, which then fuses to the central M-proton/g/ā^{M} unit's dY1r, down quark on side A'C'.

The exact same type of process as previously described takes place once more. The PP monopole and VV monopole in each of the two magnetic Z bosons present in each of the two opposite-faced dY1r units, · PP : VV · , again repel each other, due to the opposites repel rule in magnetic dimensions. Both dY1r units then become unstable due to the greater distance in place between the PP and VV monopoles. This causes the PP unit on side A"C" to cross-bond with the PP unit on the side A'C', forming a $\left(\genfrac{}{}{0pt}{}{PP}{PP}\right)$, W- boson; which then forces the two VV monopoles on each side of each PP unit to also cross-bond, creating two $\left(\genfrac{}{}{0pt}{}{VV}{VV}\right)$, W+ bosons; as well as also then causing the three remaining PP monopoles on sides A"C" and A'C' to produce three more $\left(\genfrac{}{}{0pt}{}{PP}{PP}\right)$, W- bosons, in order to establish a stable structure.

This structure is called a double-quark:
$\left[\begin{matrix} PP & PP & PP & VV & PP & VV \\ PP & PP & PP & VV & PP & VV \end{matrix}\right]$ (DQ).

Now, as a result, both the right side M-neutron/g/ā^{M} and the left side M-neutron/g/ā^{M} units are each spliced to the central M-proton/g/ā^{M}, creating a new, third isotope of the element hydrogen, M-tritium/g/ā^{M} (p.205)

Although this isotope has a half-life of approximately 12.5 years in its ionic, atomic or molecular form, in its magnetic form it is not radioactive and thus can unite with other isotopes to produce stable end products, in the form of other isotopes of other elements.

Diagram 12-III-D

M-Tritium/g/$\bar{a}^{M}$

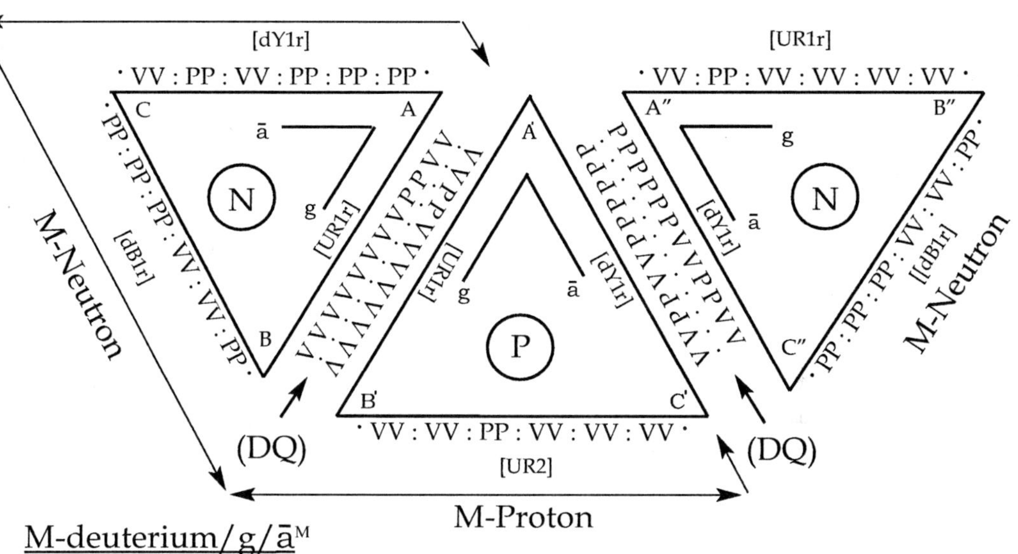

Diagram 12-III-E

Tritium/g/$\bar{a}^{(+)1}$

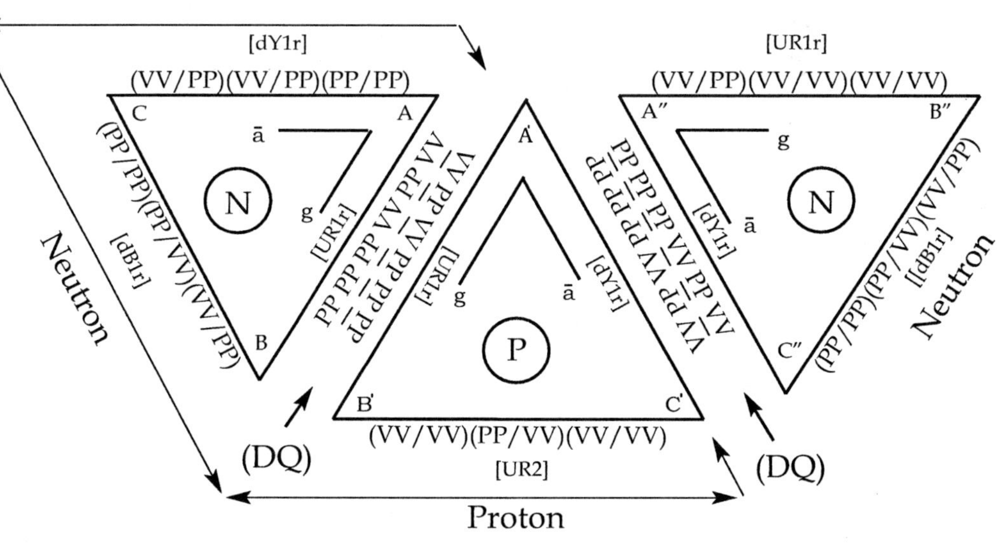

Deuterium/g/$\bar{a}^{(+)1}$

As with its two sister isotopes, the proton/g/ā^{M} (hydrogen) and deuterium/g/ā^{M}, it transforms into an electric form below 10 million K. The tritium/g/ā$^{(+)1}$ ion can be seen in Diagram 12-III-E, p.205. It will eventually becomes a neutral atom with the addition of a (-)1 electron. Then, it covalently bonds with a like tritium/g/ā atom to form the tritium molecule, $({}^{3}_{1}H)^{0}_{2}$; which, along with $({}^{1}_{1}H)^{0}_{2}$ and $({}^{2}_{1}H)^{0}_{2}$, become the ingredients of the first of seven stages in the evolution of our present universe and solar system.

Helium 3 and 4

The element helium can occur in five possible isotope forms: 3, 4, 5, 6 and 7; of which only two, helium 3^{0}, $({}^{3}_{2}He)^{0}$, and helium 4^{0}, $({}^{4}_{2}He)^{0}$, are not subject to radioactive decay, and are therefore stable.

The reaction leading to the synthesis of helium 3^{0}, is shown in Diagram 13-M, below, as follows.

The deuterium/g/ā^{M} (deuterium^{M}), the synthesis of which has been previously described and presented in Diagram 11-M-D, p.201, has its unreacted dY1r, down quark, component of its

Diagram 13-M

Helium 3^{M} , $({}^{3}_{2}He)^{M}$

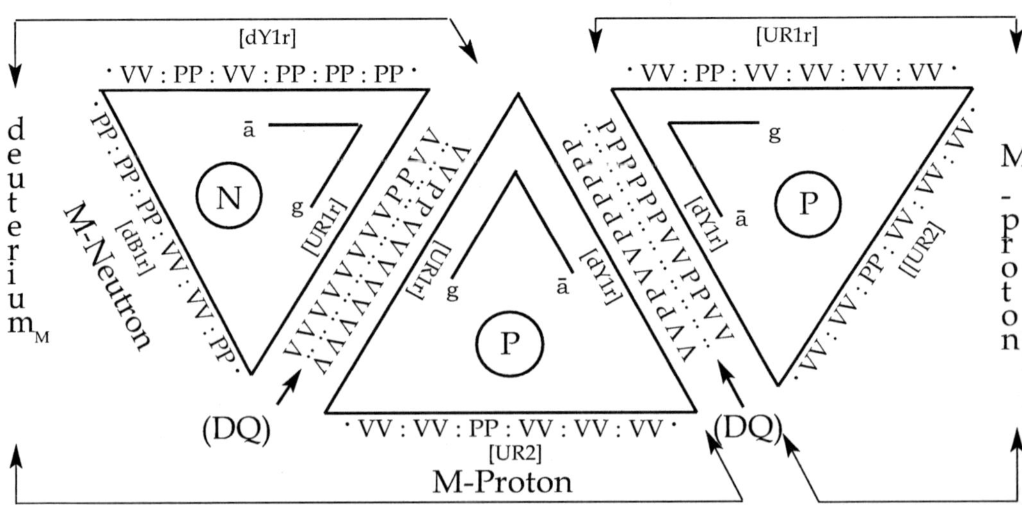

Diagram 13-E

Helium $3^{(+)2}$, $({}^{3}_{2}\text{He})^{(+)2}$

+]2

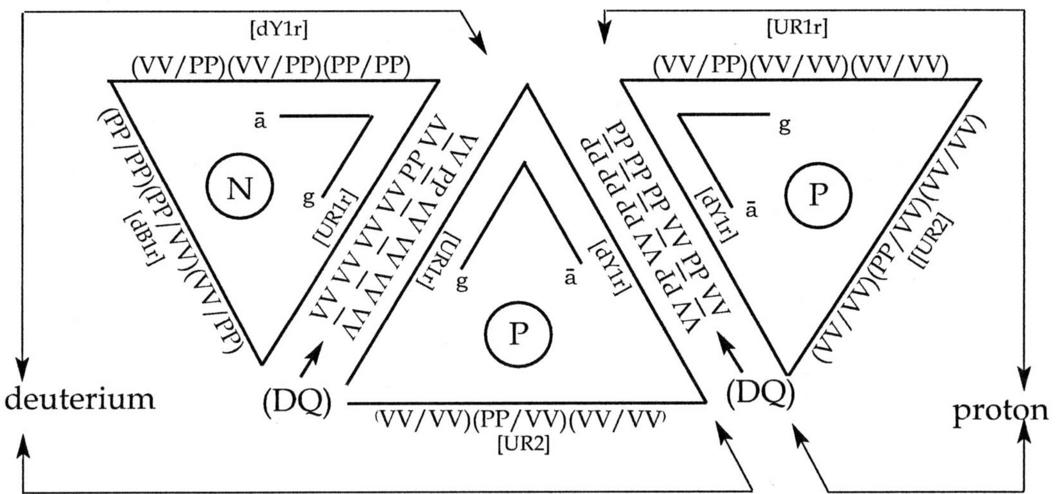

M-proton, (P), impacted by another dY1r, down quark which is an unreacted component of another single, free M-proton, (P).

Both of these two dY1r components are then fused together into a "double-quark"; which also splices together their other attached components to produce helium 3^{M}, $({}^{3}_{2}\text{He})^{M}$ (Diagram 13-M, p.206,) using the exact same process as previously described for the fusion of two other dY1r components: one in an M-proton section of a deuterium^{M}; the other in a single, free M-neutron, to create tritium^{M}, as shown in Diagram 12-III-D,p.205. Below 10 million K, He 3^{M} transforms into its electric (ionic) form, as seen in diagram 13-E, above.

Helium 4, $({}^{4}_{2}\text{He})^{0}$, also employs this same fusion process. As seen in Diagram 14-M, p.208, two same deuterium^{M} units strike each other at and between their UR2, up quark, components of each of their M-protons (P).

This action converts each of their three horizontal "intraquark" bosons within each of their two opposite positioned UR2, up quarks, into six "interquark" bosons, now positioned vertically between the two former horizontal UR2. This configuration is now referred to as a "double-quark" structure. It acts to splice together

Diagram 14-M

Helium 4^M, $({}^{4}_{2}\text{He})^M$

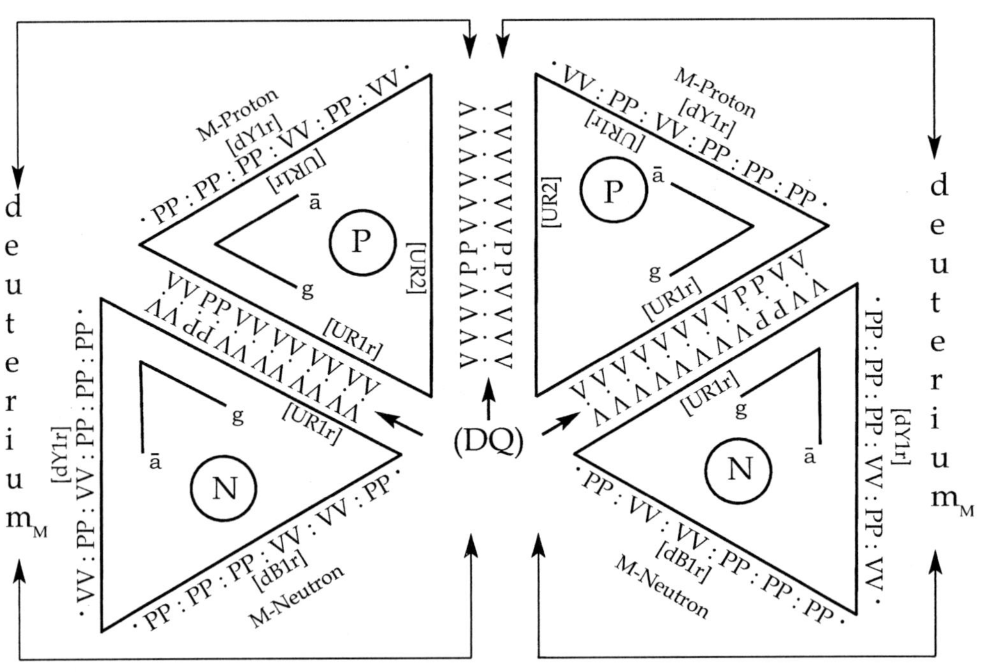

Diagram 14-E

Helium 4^{0}, $({}^{4}_{2}\mathrm{He})^{(+)2}$

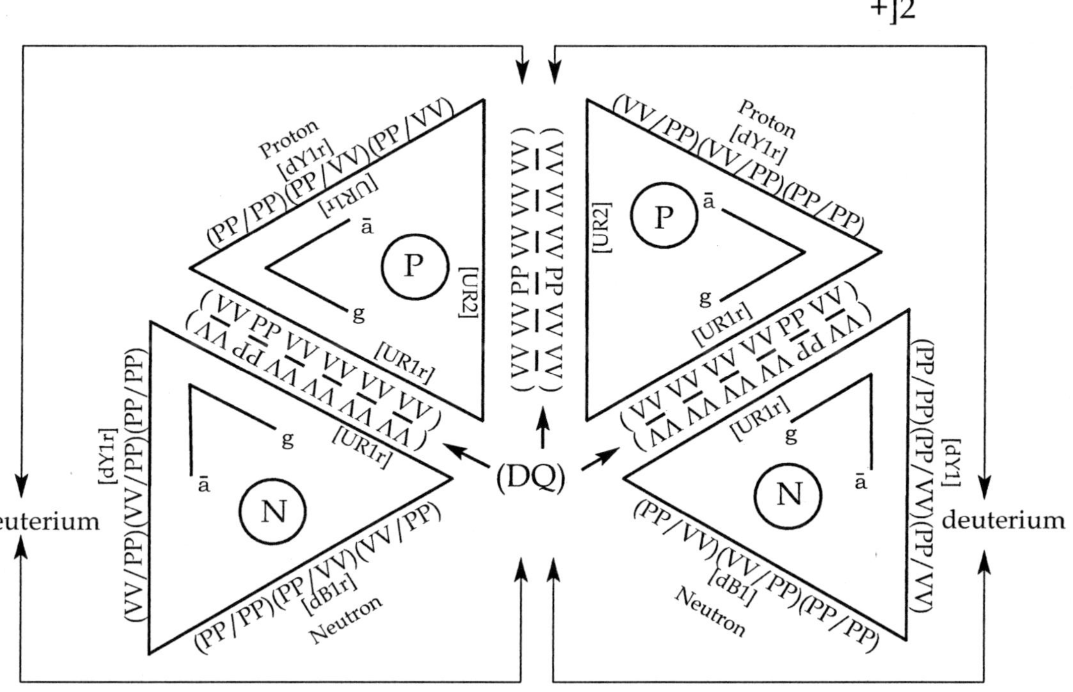

and fuse the two magnetic deuterium^{M} units into the helium 4^{M} isotope.

After this new isotope's temperature falls below 10 million K., it will appear, as illustrated in Diagram 14-E, above, as an ion, which then picks up two electrons to produce a neutral helium 4 atom. The helium 3 ion is depicted in Diagram 13-E, p.207.

The above described fusion process which produces all of the isotopes of all of the elements of the periodic table, except the proton and neutron, is therefore universal: two identical quarks (up or down), whether components of two different, single M-protons; or an M-proton and an M-neutron; or two different single M-neutrons; must initially impact each other.

Then each of the three horizontal "intraquark" boson components within each of the two impacting and opposite positioned quarks are then cross-bonded into six vertical positioned "interquark" bosons between them: which combined structure is now classified as a single double-quark (DQ).

The six vertical bosons, now between the two impacting quarks, referenced as a "double-quark", splice and fuse all entities attached to it into a new, single entity, a new isotope of an element.

Lithium 6

The element lithium can appear as any one of five different isotopes: 5, 6, 7, 8 or 9. However, only lithium 6 and 7 are stable; while 5, 8 and 9 are radioactive and decay.

In Diagram 15-M.1, p.213, we depict a magnetic helium 4^M $({}^{4}_{2}He)^M$ and a magnetic deuterium^M $({}^{2}_{1}H)^M$ bonded and fused together, into a lithium 6^M unit. It was created by the impact of a horizontal dY1r in a proton (P) in the helium 4^M upon a dY1r in a proton (P) in the deuterium^M at the bond (B5) site. This produced six vertical "interquark" bosons between the two horizontal dY1r, down quarks, of the two protons (P).

The six vertical "interquark" bosons are a unity, identified as a double-quark, which are now vertically cross-linked, instead of being two horizontal quarks, each containing three "intraquark" bosons, both of which are not cross-linked.

The magnetic bond between each of the two monopoles which thereby creates each of the six "interquark" bosons in the double-quark is seen as a · · . This · · symbol is also found in four additional double-quark structures present within the lithium 6^M unit: at (B2) (proton to proton); and (B1), (B3) and (B4) (proton to neutron). It identifies these four, double-quark units, as having been formed previously in the building process which created the individual helium 4^M and deuterium^M units, which were then fused together to produce the lithium 6^M in Diagram 15-M.1, p.214.

Another symbol, : , is also found in the lithium 6^M diagram. It signifies and identifies a magnetic bond as being a magnetic "intraquark" bond, which in turn produces an "intraquark" boson, one of three always present in a horizontal magnetic up or down quark .

The use of the two different symbols for the magnetic bond between monopoles, both · · and : being the exact same type of bond, is to visually simplify for the reader the recognition of a double-quark, with its compliment of vertical six · · bonds between two rows of monopoles; in contradistinction to the single quark, with its tell-tale : bonds within its three horizontal bosons.

When the extant temperature causes the lithium 6^M unit to fall below 10 million K., the magnetic structure and its bonds are converted into an electric, or ionic, structure and bonds, as illustrated in Diagram 15-E.1, p.214.

In this diagram, each of the · · magnetic bonds between each of the six vertical "interquark" bosons, in the double-quark structure, are replaced by an electric bond, symbolized by a horizontal line, —;while the : magnetic bond in each of the three horizontal "intraquark" boson in a single quark is substituted by a slant line, /. Both of these two line symbols are exactly the same electric bond, but are used to differentiate visually between a (-) double-quark and (/) single-quark structure.

Additionally , each single dot, · , at the end of each magnetic quark, becomes a) or a (in an electric quark. When a single) or (is found in a quark, the nearest) or (in another quark would create a boson-to-boson bond between them. There are three of these bonds in a proton or neutron,which holds each together.

Lithium 7

Diagram 16-M.1,p.215, illustrates the synthesis of magnetic lithium 7^M (${}^{7}_{3}\text{Li}^{M}$) ; bonding together a helium 4^M (${}^{4}_{2}\text{He}^{M}$) and a tritium^M ($^3_1H^M$). It is achieved using the same universal method previously described: converting the three horizontal "intraquark" bosons in two impacting and contiguous quarks in each of the two entities into six vertical "interquark" bosons between the two impacting quarks. The vertical six bosons are classified as a double-quark (DQ).

In Diagram 16-M-1, p.215, it is two dY1r, down quarks, one in each of two protons (P) located at bond B4 , which convert each of their three horizontal "intraquark" bosons into the six "interquark" vertical bosons between them, thereby fusing together the helium 4^M and tritium^M into lithium 7^M. The new lithium 7^M has five additional six boson, vertical double-quark bonds present, which were formed previously in the construction of the helium 4^M and tritium^M units: a single proton-to-proton double-quark bond at (B2) ; and four, single proton-to-neutron double-quarks at (B1), (B3), (B5) and (B6) . All of these double-quark bonds (DQ) were created by using the same universal "intraquark" conversion to "interquark" method, as previously described.

In fact, the synthesis of every isotope of every element in the periodic table, except the proton and neutron, has used this above described "universal" method in its construction. This includes the neutron-to-neutron conversion to a double-quark structure which is used in the construction of isotopes of heavier elements.

As the temperature of the lithium 7^M dips below 10 million K., the entire magnetic structure and its magnetic bonds convert into an electric or ionic structure composed of electric bonds. This is depicted in Diagram 16-E.1, p.216.

Lithium 6^M can also be synthesized by the double-quark bonding of helium 3^M and tritium^M; while lithium 7^M can be made by the double-quark bonding of a neutron to a helium 6^M unit.

Each isotope has several paths to its synthesis. As the isotopes increasingly add protons and neutrons to their structure, the number of paths leading to their synthesis also increases. This is true for stable isotopes as well as radioactive ones.

Diagram 15-M.1

Lithium 6^M $({}^6_3Li)^M$

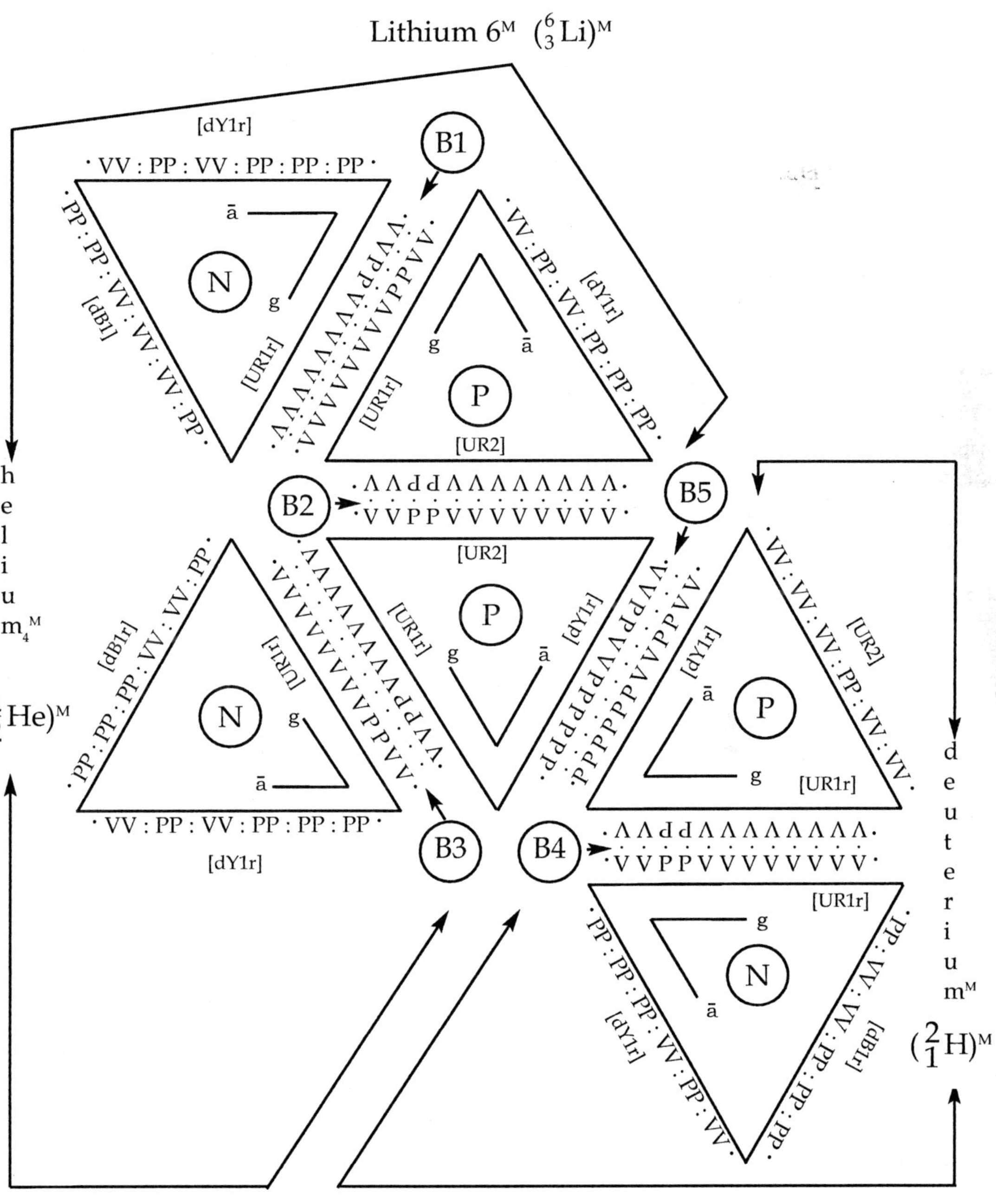

Note: All (B) bonds are double-quark bonds.

Diagram 15-E.1

Lithium $6^{(+)1}$ $({}^{6}_{3}Li)^{(+)1}$

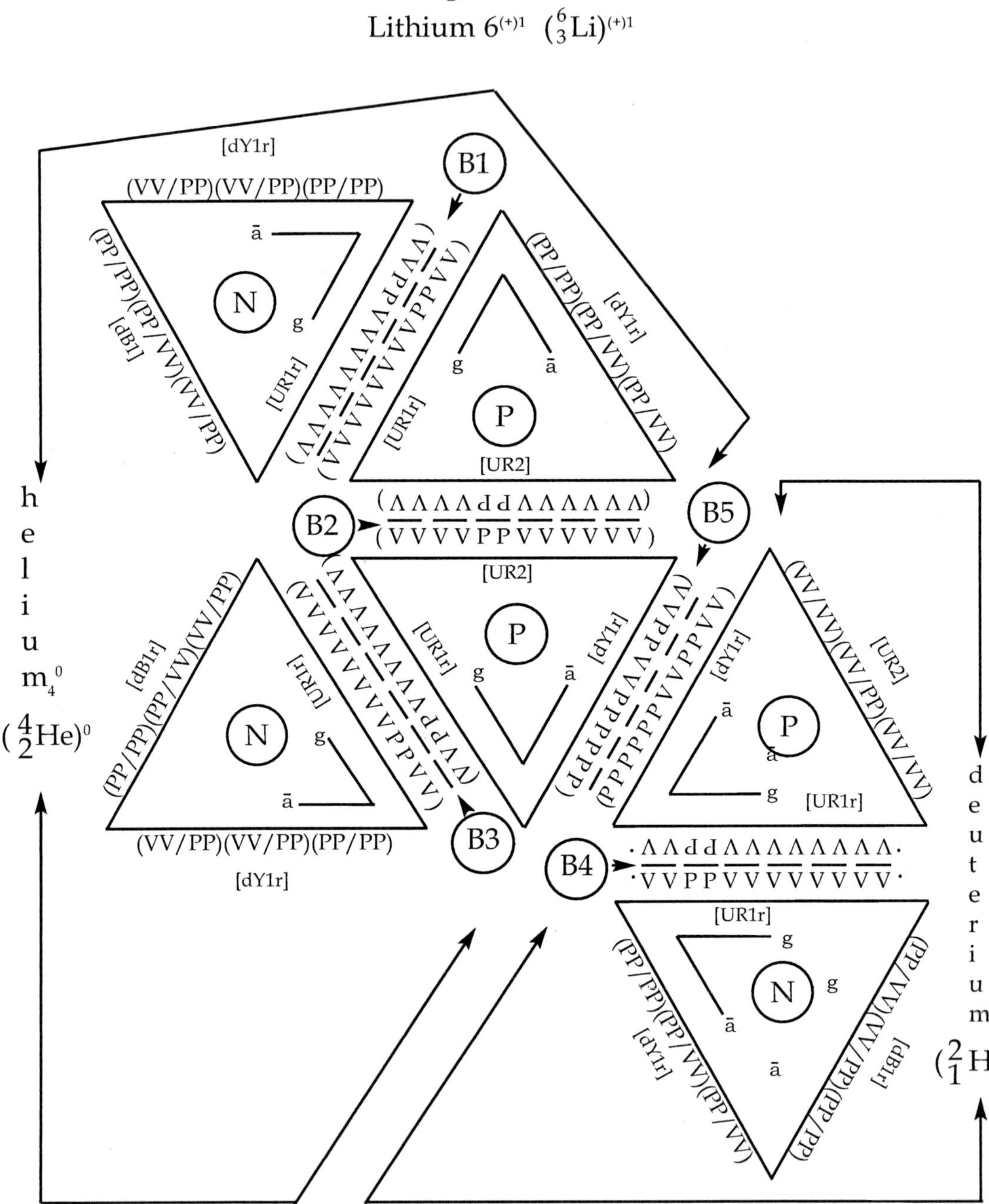

Note: All (B) bonds are double-quark bonds.

Diagram 16-M.1

Lithium 7^M, $({}^{7}_{3}\text{Li})^M$

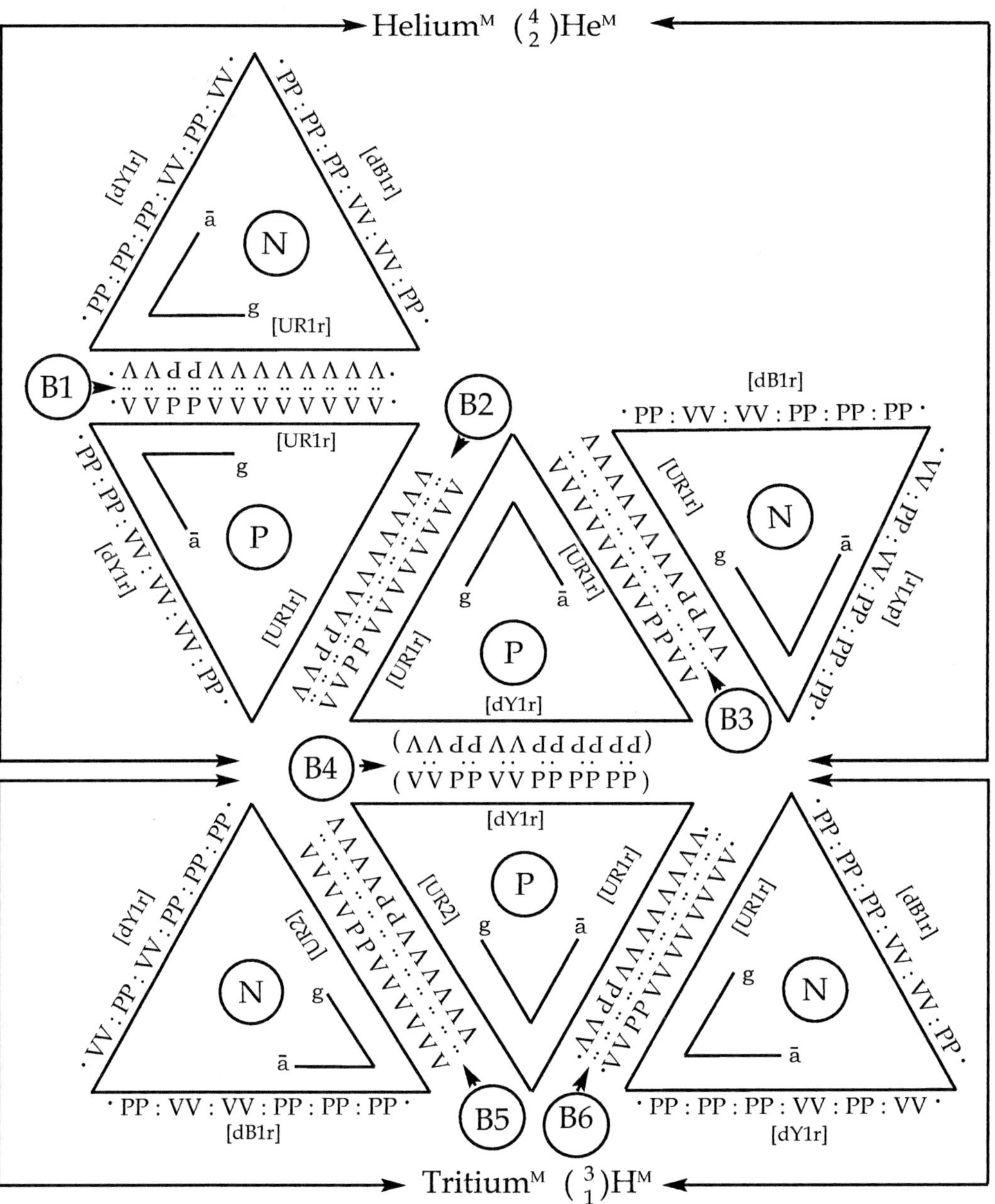

Note: All (B) bonds are double-quark bonds.

Diagram 16-E.1

Lithium $7^{(+)}$, $({}^{7}_{3}Li)^{(+)1}$

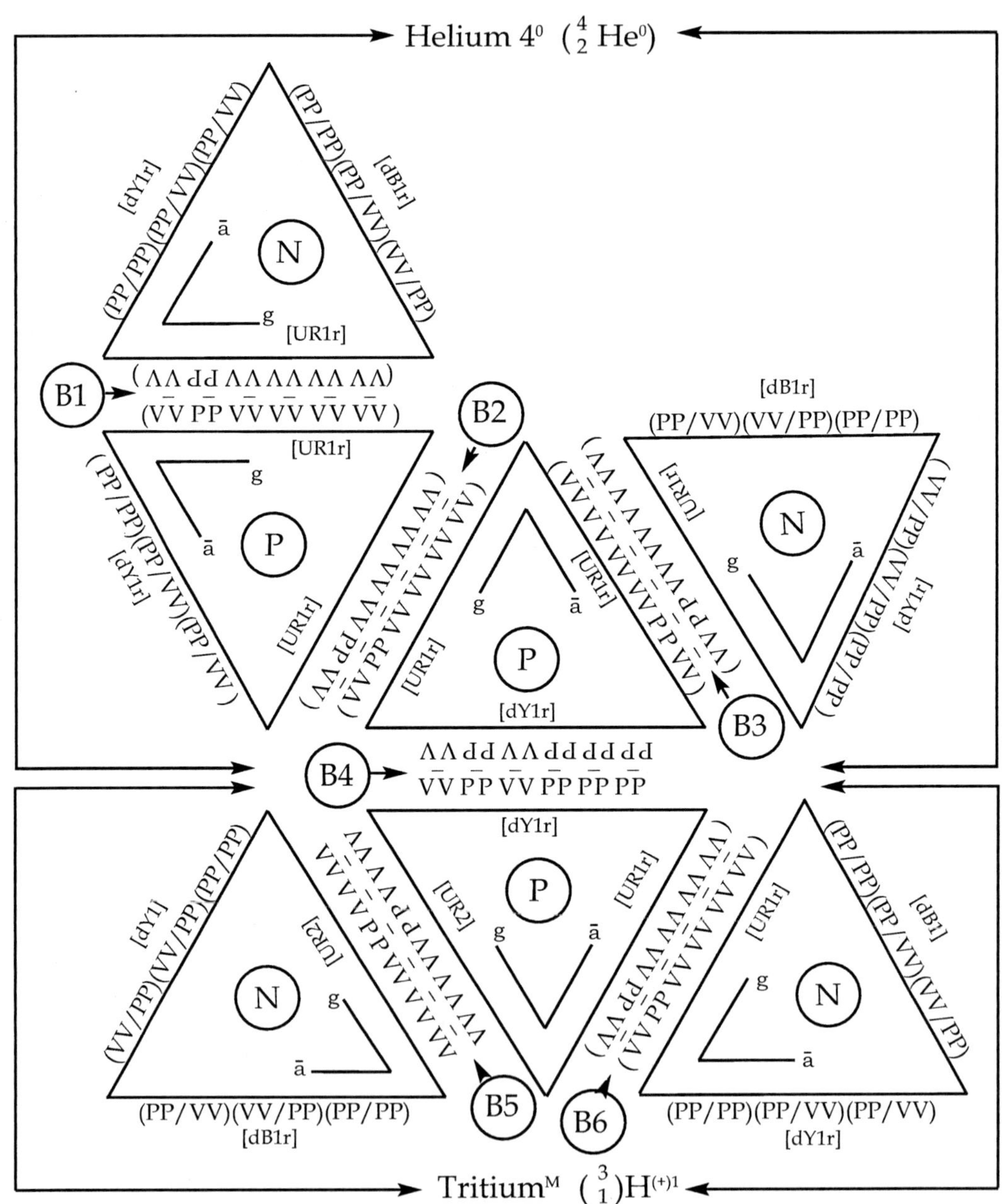

Note: All (B) bonds are double-quark bonds.

Equivalent Double-Quark Isomers

In diagram 15-M-1, P.213, we note that the dY1r to dY1r, double-quark, at site B5 , spilcing together the helium 4^M and the deuteruim ^{M}, to create lithium 6^M, is actually only one of four such dY1r sites in the helium 4^M which could bond to either of two dY1r sites in the deuterium.

The other three dY1r sites in helium 4^M are: one in the second proton Ⓟ, adjacent to the first proton Ⓟ at B5 ; and one in each of the two neutron Ⓜ present.

The second of the two dY1r sites in the deuterium ^{M}, in addition to the first site in the Proton Ⓟ at B5, is in the neutron Ⓝ component.

Thus, there are eight possible equivalent dY1r,to dY1r, double-quark, which can be formed to produce lithium 6. Each would be an isomer of the lithium 6 isotope.

Yet; only one type of lithium 6 isotope has been identified to date.

This is counterintuitive, since in organic chemisty each of these double-quark site units would constitute a seperate isomer, capable of being isolated.

In certain isotopes of heavier elements, however, we do find the phenomenom of a single radiactive isotope displaying several half-lives.

This indicates the presence of equivalent isomers; similar to those indicated above as examples of lithium 6 equivalent isomers.

Seven Stages of Element Synthesis

We have previously illustrated and discussed the steps necessary for the synthesis of the first three elements in the periodic table: hydrogen, helium and lithium.

The element Hydrogen is created in the form of three isotopes: hydrogen ($^1_1$H), deuterium ($^2_1$H) and tritium ($^3_1$H) . These three then become the feed-stocks for Stage 1, the first of seven stages of element synthesis in the universe, as designated in

Diagram 17, below. Each of the first six stages ends its life cycle in a supernova event (Stages 1, 2, 3, 4, 5 and 6). In the case of Stage 1, two new elements, helium and lithium (He-Li), are produced in the supernova event (as seen under column, "No. of Elements Supernova Adds").

These two elements, plus hydrogen, in its three isotope forms, brings the end total number of elements present to three; which is also the number listed under the "End Total No." column in Diagram 17 for Stage 1.

These three elements, in turn, become the feed-stocks which begin the life cycle of Stage 2, where the same number 3 is listed under the column "Beginning No. of Elements". Thus, the "End Total No." column of the previous stage is always the same number (and same elements) with which the next stage begins its new life cycle.

Diagram 17
Seven Stages of Element Synthesis

Stage or World No.	Beginning No. of Elements	Beginning No. of Elements Symbols	No. of Elements Supernova Adds	End Total No. of Elements
1	1	H	2 (He-Li)	3
2	3	H-Li	3 (Be-C)	6
3	6	H-C	6 (N-Mg)	12
4	12	H-Mg	12 (Al-Cr)	24
5	24	H-Cr	24 (Mn-Cd)	48
6	48	H-Cd	48 (In-Cm)	96
7	96	H-Cm	———	———

Therefore, in Diagram 17, Stage 2 begins with three elements, H-Li (hydrogen through lithium), the same number and same elements with which Stage 1 ended at the conclusion of its supernova event ("End Total No.").

Stage 2 then concludes its life cycles' supernova event with the synthesis of three new elements, Be-C (beryllium through carbon), which, when added to the three elements (H-Li) it started with, produce an "End Total No." of 6; which is also the same number and same elements Stage 3 initiates its life cycle (H-C).

Stage 3 then progresses, finishing with its supernova event by adding six new elements, N-Mg (nitrogen through magnesium), for an "End Total No." of 12 elements present.

These same 12 elements then initiate Stage 4 (H-Mg), which also ends its life cycle with a supernova event and adds 12 new elements, Al-Cr (aluminum through chromium), for an "End Total No." of 24 elements present; the same number and same elements with which Stage 5 commences (H-Cr).

Stage 5, in turn, ends its life cycle with a supernova event, which adds 24 new elements, Mn-Cd (manganese through cadmium), for an "End Total No." of 48 elements.

These same 48 elements then start the life cycle of Stage 6, H-Cd; which then produces 48 new elements in the fires of its concluding supernova event, In-Cm (indium through curium), for a combined "End Total No." of 96 elements.

The explosive conclusion of Stage 6 sets in motion a gas cloud, which initiates Stage 7. This gas cloud is also the initial stage in the evolution of our present solar system.

The 96 elements which composed the gas cloud have, since that time, had their numbers diminished through the radioactive decay of all the transuranium elements, Pu-Cm, (plutonium through curium), as well as such elements as technetium (Te).

All of the foregoing six stages, can be seen as a thumbnail sketch of a 10 billion year transformational history. However, the complexity inherent and the myriad deviations from the overall core description, illustrated in Diagram 17, can only be definitively explained and resolved in a future work.

Another insight not previously discussed concerns itself with the fact that nothing in our existent universe is absolute.

In Diagram 17, we list under the column, "No. of Elements Supernova Adds", the specific numbers of elements and which specific ones are actually added. For example, in Stage 1, two additional elements are added, He-Li, helium and lithium; while in Stage 4, 12 additional elements are added through a similar supernova synthesis, and these are Al-Cr, aluminum through chromium in the periodic table.

Yet, up to approximately 5% of the total number of atoms present at the end of each supernova event listed in Stages 1 through 4 are composed of elements of the periodic table that come just after the last element listed in the column, "No. of Elements Supernova Adds", up to and including the element nickel. For example, the last element in that Stage 1 column is lithium. Therefore, 5% of the total of all of the elements present at the end of Stage 1 would be beryllium through nickel; while 5% of the total at the end of Stages 2, 3 and 4 would be nitrogen through nickel, aluminum through nickel and manganese through nickel, respectively.

Each of the four groups, just cited as being present up to an anomalous 5% of the total quantity of elements produced in the four supernova events, are a direct consequence of not being able to completely shut down the supernova event instantaneously. This will be discussed in greater detail in the "Step-by-Step Element Synthesis" section.

The iron and nickel atoms, which are present at the beginning of Stages 2, 3, 4, 5, 6 and 7, are responsible for the organization of the gas cloud that results from the explosive supernova event which ended each of the previous six stages (1-6). The quantities of iron and nickel produced by the end of Stages 1, 2, 3 and 4 are very small, yet are still able to effectively organize the resultant gas cloud which inaugurates the next stage in evolutionary development.

The iron and nickel atoms are magnetically attractive and spin in a counterclockwise direction. Cobalt, the third and final member of this ferromagnetic group, has the former property, but not the latter. Also it is present in relatively small quantities.

The iron and nickel atoms' counterclockwise spin causes the other isotopes of elements present in the gas cloud to move in the same direction; as well as also initiating a magnetic-based attraction and agglomeration of all of the ferromagnetic elements, which then serve as large-body foci upon which other non-ferromagnetic elements can be accreted.

These ferromagnetic-centered, counterclockwise spinning bodies eventually build up through a series of impacts and mergers; concluding in the organization of a sun and a number of planets.

This can be surmised from the composition of our own sun, which, in addition to its approximately 73.4% hydrogen and 25% helium, also contains 0.12% ferromagnetic element content (all of the foregoing being percentages by mass). This latter percentage of the enormous mass of the sun (1.989×10^{33} g.) is quite sizeable and is thus is able to organize the entire initial gas cloud resulting from the 6th stage supernova event into our present sun and it's planets.

Although relatively small in comparison to the hydrogen and helium content of the sun, these iron and nickel atoms act as solar catalysts in organizing the massive number of other atoms present in the sun into its present shape, as well as organizing other atoms into the planets and moons.

These ferromagnetic elements also act as primary catalysts in many other reactions. It appears to be their underlying nature. We will comment further on the role of these ferromagnetic elements in the section on the organization of our present solar system.

Step-by-Step Element Synthesis

The seven stage plan of element synthesis, outlined in Diagram 17, is underpinned by a step-by-step process made necessary by the numerous possible color quark combinations inherent in the structure of every isotope of every element in the periodic table. Another contributing factor is the effective concentrations of the reactant isotopes present in order to effectuate the process of building a new and heavier isotope out of smaller and lighter isotopes; all of this, off course, occurring within the time constraints imposed by a supernova event, whose time of existence is directly related to the solar mass of a sun. The greater the solar mass, the shorter is the time of the supernova event.

Each of the many isotopes of each element can be constructed from twelve color types of down quarks: 6 yellow, 5 blue and 1 red; and six color types of up quarks: 4 red and 2 yellow.

The gluon/antigluon meson does not directly take part in the synthesis reaction of lighter into heavier elements, other than in a decay of a proton or neutron within the structure of an isotope, which then changes the types of color quarks available in an isotope which is about to undergo a synthesis reaction and transformation.

Additionally, there are many routes of reaction in creating a single isotope of an element; all of them requiring that the exact same color quark on two separate isotopes strike each other directly in order for their two isotopes to be able to merge and thereby form a heavier isotope of a new element. Although the foregoing is complex in nature, it is quite decipherable and doable.

For example, in order to synthesize the two stable isotopes of lithium: 6 and 7 ($^{6}_{3}Li$ and $^{7}_{3}Li$), we first have to generate helium 4 ($^{4}_{2}He$), deuterium ($^{2}_{1}H$) and tritium ($^{3}_{1}H$). Thereafter, $^{4}_{2}He$ and $^{2}_{1}H$, as well as $^{4}_{2}He$ and $^{3}_{1}H$, combine under magnetic dimensional conditions to respectively create $^{6}_{3}Li$ and $^{7}_{3}Li$.

Other routes of synthesis, including one involving helium 3 ($^{3}_{2}He$) and tritium ($^{3}_{1}H$) can also produce lithium 6 ($^{6}_{3}Li$).

Lithium 7 ($^{7}_{3}Li$) can additionally be formed through a route that first forms the unstable helium 6 ($^{6}_{2}He$) isotope, and then adds hydrogen ($^{1}_{1}H$). All of the foregoing routes are under magnetic dimensional conditions (10 million K. plus).

In the latter case, even though one of the two feed-stocks was the unstable ($^{6}_{2}He$), the resulting isotope ($^{7}_{3}Li$) will remain stable; since it is not the feed-stock which determines stability in electric or neutral-electric dimensions but only the end product; which, in this case, is the stable lithium isotope ($^{7}_{3}Li$).

The usually unstable (radioactive) electric phase beryllium isotope $^{8}_{4}Be$, is quite stable when created in a magnetic phase dimension. These magnetic phase reactions are illustrated as follows:

1. $^{4}_{2}He + ^{4}_{2}He \rightarrow ^{8}_{4}Be$

2. $^{7}_{3}Li + ^{1}_{1}H \rightarrow ^{8}_{4}Be$

3. $^{6}_{3}Li + ^{2}_{1}H \rightarrow ^{8}_{4}Be$

4. $^{5}_{3}Li + ^{3}_{1}H \rightarrow ^{8}_{4}Be$

The end product of these four possible magnetic routes, $^{8}_{4}Be$, can then be converted into the stable carbon isotope $^{12}_{6}C$, as follows:
5. $^{8}_{4}Be + ^{4}_{2}He \rightarrow ^{12}_{6}C$

There are numerous other magnetic step-by-step routes to $^{12}_{6}C$, some of which are as follows:

6. $^{6}_{3}Li + ^{6}_{3}Li \rightarrow ^{12}_{6}C$

7. $^{7}_{3}Li + ^{5}_{3}Li \rightarrow ^{12}_{6}C$

8. $^{9}_{4}Be + ^{3}_{2}He \rightarrow ^{12}_{6}C$

9. $^{11}_{5}B + ^{1}_{1}H \rightarrow ^{12}_{6}C$

10. $^{10}_{5}B + ^{2}_{1}H \rightarrow ^{12}_{6}C$

11. $^{9}_{5}B + ^{3}_{1}H \rightarrow ^{12}_{6}C$

All of the foregoing lithium isotopes are created in the supernova event at the end of Stage 1, as exhibited in Diagram 17.

The reactions creating beryllium (Be), boron (B) and carbon (C) occur mainly during the supernova event ending Stage 2.

Underlying all of the isotope building reactions, in the supernova events ending stages, 3, 4, 5 and 6, are the difficulties outlined at the beginning of this chapter; in particular, having to have the exact same type of color quark on two separate isotopes strike each other directly in order to create a double-quark, which initiates the merger process that then creates a new and heavier isotope out of two lighter ones.

If two different types of color quarks on two separate isotopes strike each other directly, no merger process ensues; and the two isotopes bounce off each other to seek out other isotopes with the right type of color quark. Here, the laws of probability determine how many hits must take place before two same types of color quarks can strike each other and effect merger into a new and heavier isotope.

Factors affecting the above probability are: the concentration of reactant atoms and the time required to produce them. For example, in the Stage 2 supernova event, the concentration of the beryllium isotope, ${}^{8}_{4}Be$, must be high enough in order that one of possibly 12 color types of down quarks which may be present in its structure can find and strike the exact same color type of down quark, also one of 12 types possible, in a helium ${}^{4}_{2}He$ atom, which also must be present in sufficient quantities to produce ${}^{12}_{6}C$.

Then, the same high concentrations of ${}^{12}_{6}C$ must be present in order to produce elements in the periodic table beyond ${}^{12}_{6}C$, up to and including nickel. Yet, only 5% of the total present at the close of the supernova event are such "beyond" elements.

Obviously, the concentration of necessary isotopes of elements and the time necessary for their buildup and reaction was so limited as to produce only up to 5% of the total.

Stages 3 and 4 also suffer from these same constraints in each of their supernova events.

In Stages 5 and 6, the above constraints of concentration and time span of the supernova event are further complicated by the presence of new isotopes which generate large numbers of neutrons for capture to produce new and heavier elements in the periodic table, as seen in Diagram 17.

Yet, with all of the constraints on the synthesis of all of these elements in the periodic table, our own nascent solar system eventually did appear, after the completion of six stages, with 96 elements as the initial number for beginning the seventh stage of evolutionary development. The six previous stages had spanned approximately 10 billion years in total.

The Evolution of the Solar System

A gas cloud, composed of a heterogenous dispersion of approximately 96 elements, hydrogen through curium, which is a remnant of an explosive Stage 6 supernova event, sets out on a Stage 7 evolutionary path; which, in approximately 4.7 billion years, will result in our present solar system, an N11 dimension.

The gaseous iron and nickel components of this irregular-shaped spherical cloud immediately begin to spin in a counterclockwise direction; as well as concurrently initiating a coalescing process between themselves, based on their inherent magnetic property. All other elements, except cobalt, lack this property. The two, plus cobalt, then further cool, passing through a liquid state into a solid state.

The powerful impetus, directed by these two ferromagnetic elements' counterclockwise spin, has propeled all of the other elements in the cloud in the same direction.

This directed motion, secondarily, will cause the initial spherical-shaped cloud into developing an organizing process, which over a span of time, changes its initial configuration into that of a disc-shape, which is divided into 10 distinct sectors.

The middle, first sector, of the evolving disc is a massive hydrogen-rich sphere. It is the protosun of the system.

The iron/nickel components of this first sector will endlessly sweep through it, some growing in size to predominantly that of grains of sand or, in certain instances, to that of giant bolides, approximately a moon in size, as well as a myriad number of other bodies somewhere between these sizes.

In each of the next eight sectors, radiating out from and orbiting the middle protosun sector, the iron/nickel components also grow in size, adding numerous other types of accreting elements to that of their initial minute size grains, which then begin to smash into each other, initiating a merger process between themselves which produces ever larger bodies. These are then structurally built up layer upon layer, due to a process of accretion that sweeps up elements and compounds found in each of the eight sectors onto these iron/nickel-core nuggets.

Upon smashing into each other, they merge their contents; the layers then mix together but always settle around a central cavity containing molten iron/nickel, which is kept in this state by the radioactive decay of various iron and nickel isotopes, as well as other radioactive isotopes including those of cobalt.

This process continues on, with ever larger bodies forming, until a single, gigantic protoplanet results in each of the eight sectors; which in order from the center protosun are: Mercury, Venus, Earth, Mars, Jupiter, Saturn, Uranus and Neptune.

The tenth and final sector, also orbiting the protosun, extends from the outer edge of the Neptune system to the outer edge of the solar system. It includes Pluto, understood herein as also being at the same time a moon of Neptune; the Kuiper Belt; and the Oort Cloud.

Each of the eight gigantic protoplanets in their formation have swept up gaseous, liquid and solid elements and compounds extant in their sector, dispersing them throughout the myriad moving and churning slurry-type layers they contain; which, in turn, surround a central cavity of molten iron and nickel, the center of which is a solid iron/nickel core under great pressure. The iron and nickel components of the central core cavity are the foundation of each protoplanet and an N10 dimension.

The amount of radioactive isotopes each of the eight protoplanets has acquired during its myriad sweeps of its sector is also a determining factor in how long it can maintain its uppermost layers and surface area sufficiently hot enough to continue its ever-moving slurry state, thus preventing a change of state into a cooler type of gel state; which is differentiated from a slurry state by the complete cessation of all movement in a layer or surface area.

While the eight protoplanets are each metamorphosing, the hydrogen-rich protosun has been developing a multilayered structure of its own, at the center of which is a relatively small core-area of almost pure hydrogen gas. It becomes the subject of ever-increasing pressure and temperature, resulting from the organizing process, previously described. As this process continues, there is a gradual contraction of this hydrogen core-area, leading initially to a portion of it igniting and burning (non-fusion).

This ongoing reaction further increases the temperature and pressure of the hydrogen in the entire core-area, until 10 million K. is exceeded; at which point most of the core-area hydrogen suddenly fuses into helium, causing an enormous explosion, which then blows out nearly all of the solidified iron/nickel accretion bodies present in the rest of what is now a new sun, including the many large iron/nickel accretion bodies resting in an area at the very edge of the sun sector. These bodies then shoot out across the disc's other eight sectors, many of which bodies then collide with the protoplanets therein, as well as some being deposited in what will be the asteroid belt between Mars and Jupiter, the Kuiper Belt, as well as the Oort Cloud.

This single explosive event will significantly alter all of the protoplanets except the protoearth, due to the latter being the only protoplanet still in a completely slurry state at the time of the explosion.

All of the protoplanets to that point had evolved through an initial stage where almost all of the gaseous, liquid and solid elements and compounds present in their sector had been amalgamated into one gigantic planetary body: a layered structure enclosing a central cavity containing molten iron/nickel, which then enclosed a core of solid iron/nickel under immense pressure.

The upper layers of these protoplanets were in a slurry state, which was constantly moving and churning, held together in a spherical form by gravity.

All of the protoplanets, except the protoearth, had next graduated to the stage where gel producing compounds were created, which were then introduced and incorporated into the various layers, eventually resulting in the entire surface layer becoming a non-moving gel. The protoearth lagged far behind time-wise in this development, due to its high concentration of radioactive isotopes which produced prodigious amounts of heat, which prevented gel formation at the surface. It eventually did develop the gel surface layer, but only several hundred-thousand years after the great, initial fusion reaction and explosion in the protosun.

Into the chaotic, churning mixture of slurry-state layers on each protoplanet, silicon dioxide (silica) and hydrated silica derivatives of magnesium, calcium, iron, sodium, potassium and aluminum were continuously being formed and admixed into the various slurry layers present, causing the formation of a myriad number of non-moving gel type areas as these areas cooled sufficiently.

Some of these gel areas are classified as muds and clays. Others, for example, which demonstrate the consistency of quicksand, are non-moving, thixotropic mud gels, which, when energy is introduced into them (shaking), liquify and lose their gel consistency. These, however, will reform as gels if no further energy is introduced and, eventually, will evolve into hard, solid areas.

Gaseous elements and compounds are also trapped in these gel state layers by the silica and hydrated silica derivatives.

Fumed, fine particle silica, for example, has been used industrially to gel hydrocarbons (butane, isopropane, etc.), as well as elemental gases such as nitrogen in aerosol products.

Once the surface of the slurry ocean covering the protoplanet, through the above described gel creating processes, begins to produce a myriad number of stable, non-moving gel areas resembling volcanic islands, in which an outgassing process is initiated in each, by which the trapped gaseous elements and compounds can escape and create an atmospheric bubble around the protoplanet. The constituents of this bubble can include hydrogen, helium, nitrogen, neon and argon, as well as compounds such as water, which also begins to puddle on the surface.

Concurrently, the temperature of the surface slurry layers has been diminishing due to a continuing loss of radioactive isotopes which have completed their decay processes, as well as the radiation of heat into the colder surroundings of the protoplanet's sector.

The loss of heat prepares the moving slurry layers to become increasingly thicker, until they stop moving, thereby transforming completely into a gel state.

The first of the eight protoplanets to form a completely non-moving gel surface, replacing its former continuous moving and churning slurry-type surface, is Neptune. This is partially due to its position as furthest from the heat of the burning protosun (non-fusion), thus the coldest; as well as partially from having a relatively low concentration of radioactive isotopes, whose reactions heat the protoplanet's surface, thus interfering with its surface gel formation. Once the surface layers, however, had cooled sufficiently, a gel state surface would be inaugurated.

Next in order of formation of a surface gel is Uranus; followed, in turn, by Saturn and then Jupiter. All three of these protoplanets did so based on their distance relative to the protosun, as well as their radioactive isotope concentration, paralleling Neptune, as discussed above.

The four inner protoplanets eventually also formed complete non-moving gel surfaces, in the following order: Mars, Mercury, Venus and Earth.

This order of formation is primarily due to the relative concentrations of radioactive isotopes contained in each. These isotopes undergo decay processes which release great amounts of heat; which, in turn, keep the surface layers of the protoplanets sufficiently hot enough to preclude gel state formation.

Eventually, the slurry layers cool, become increasingly thicker; then, in time, the surfaces become a continuous gel sheet covering the entire planet.

However, only in three of these inner protoplanets (Mars, Mercury and Venus) does this gel process go to completion, thereby allowing the release of most of the encapsulated nitrogen and water just before the protosun suddenly undergoes the massive fusion reaction which fuses all of the hydrogen in its core-area into helium. The protoplanet Earth, however, continues to remain in its slurry state for an additional 350,000 years after the protosun's fusion reaction and accompanying great blast wave which strikes all the protoplanets.

The sudden transformation of the hydrogen into helium, due to the elevation of the temperature in the center-area of the protosun beyond 10 million K., released a gigantic energy blast wave, which then ripped through each of the seven protoplanets which had completed a surface gel, catastrophically altering each of them; but leaving the protoearth intact, still encapsulating and containing all of its gaseous elements and compounds, including nitrogen and water, due entirely to its slurry state condition during the entire blast wave sequence.

The gigantic energy blast wave, which contained particles as fine as sand, as well as moon-sized bolides, moved rapidly through each sector, tearing away the gaseous, atmospheric bubble enveloping each of the seven protoplanets, which had been built up over many eons; as well as any water on the surface. Only the nascent atmosphere in the four outer planets was partially reconstituted, while Mars, Mercury and Venus lost their atmoshperes entirely.

The nitrogen and the water which were lost, as part of the complex mixture of gases and liquids blown away, were the first respective element and compound to be outgassed as soon as a surface gel state had been established on each of the seven protoplanets. Both were almost completely exhausted from all of the layers by the outgassing process in effect on each of the seven protoplanets well before the blast wave event in the protosun.

Therefore, when each of the seven protoplanets resumed the outgassing of the remaining elements and compounds still trapped in their various gel layers after the blast wave event, only minimal amounts of nitrogen (up to 3%) and water (up to 2%) were left to be generated into a new atmosphere, or onto a surface, respectively.

The protoearth, however, still in its slurry state, was generally unaffected by the blast wave event and remained in the slurry condition for at least an additional 350,000 years before finally forming a surface gel structure.

Only at that time did the Earth begin to outgas, thereby generating an atmosphere which today contains 78% nitrogen by volume, as well as a great effusion of water; which presently covers nearly 60% of the Earth's surface and, in some places, is more than seven miles deep.

This huge percentage of nitrogen in the atmosphere and immense level of water covering the surface of the planet is unique among all of the planets. Their presence on the Earth for billions of years has also produced the unique evolution of the atmosphere and surface of the planet, as well as the nitrogen-based life forms they permitted.

The high nitrogen content of the atmosphere has also greatly moderated the vast number of potential chemical reactions on the surface of the planet. This is due to nitrogen being characterized by its great inertness as a molecule (N_2), having a triple covalent bond, which prevents its decomposition even at 3000° C. It combines with other elements with much difficulty, due in great part to a very high molecular bond energy.

The great stability of the N_2 molecule causes it to resist becoming nitrogen atoms (N), which must first develop before any compound of nitrogen can be created. It's great bond strength is indicated by its very large heat of dissociation.

Nitrogen acts as an inert blanket between the most reactive chemical entities. It does not readily support oxidation; but acts as a diluting entity, slowing the rate of reaction markedly.

Nitrogen's inertness property is used industrially to increase the shelf-life of many products; often up to 1000%, by, for example, introducing a 100% nitrogen gas blanket into the space between the surface of a product and its cap; thereby displacing the pocket of air present, which contains only 78% nitrogen.

The water covering the surface of the Earth acts as an ionizing medium and a solvent for dissolving a great number of chemical entities; thus allowing them to display a valence number. This permits the formation of chemical bonds, either ionic or covalent, and their resultant compounds.

The compounds are embodied in the N7 dimension; while the valence of their components, which underpins them, are displayed as E11 dimensional electric numbers. The chemical bonds joining these components in forming a compound are assigned to the B9 dimension.

Presently, Mars and Venus have atmospheres of approximately 95% carbon dioxide (CO_2); with nitrogen present at 2.7% and 3.5%, respectively.

The Earth, in comparison, has a fraction of 1% CO_2, as well as 78% nitrogen present. It is, therefore, quite apparent that the Martian and Venusian levels of CO_2 and nitrogen, relative to the Earth's, bear out the benefits of high levels of nitrogen and low levels of CO_2 in facilitating life forms, in that Mars and Venus are devoid of life forms; whereas, Earth abounds.

Water, of course, is also a significant difference; as both Mars and Venus are now nearly devoid of water, while Earth overflows with it, as well as nitrogen-based life forms containing water.

It was the slurry state of the protoplanet Earth at the time of the initial fusion of hydrogen into helium in the protosun, which ensured Earth's uniqueness in terms of being able to bear organic life instead of carbon dioxide.

Epilogue

The foregoing journey, which flowed from the oneness and nothingness of the Singularity, via the partition event, began with the $[V_E]G$ and the $[P_E]G$ gravitons, its only deputies. These were paired in a locked embrace as V_EP_E gravitons $[(V_EP_E)G]$, thereby establishing the existent universe by their being (G1).

Thereafter, during the pangs of their separation from each other (G4), each of the two types of gravitons would create from its own unique substance its equally unique progeny, the V and the P photons, respectively; as well as the M1 dimension.

These two types of photonic raw materials would thereafter be united into a myriad number of combinations, by different processes, in each of forty-two dimensions, to eventually generate the present N11 universe; which is the fifty-fifth dimension in the complete development of the existent universe if we add the initial eleven graviton dimensions (G1-11) and the bond B1-B2.

During the course of the evolvement of the eleven dimensions of the neutral-electric phase (N1-11), six successive stages occured in which each stage concluded with a supernova event. The sixth and last such supernova event produces a total of ninety-six elements with which a seventh stage then began its evolvement.

During the subsequent development of this seventh stage, the planet earth emerged in the fourth of a ten sector solar system. The first and center sector of this system is the protosun, a predominately hydrogen gas unit, which, in its central zone, as a fully developed sun, will generate a continuous fusion reaction. This action causes the radiation of endless energy units necessary for the creation and sustaining of sentient life-forms on earth.

Sentient, herein, is defined as being capable of perceiving and responding to stimuli: light, temperature, gravity, magnetism, electric fluxes, pressure, bonds, chemical structures, etc. All of the elements in the periodic table and their respective compounds perceive and respond to at least some of the aforementioned stimuli.

The most important sentient compounds, however, are those based on nitrogen; which may additionally incorporate carbon, hydrogen and oxygen as structural elements. Some of these compounds then form derivatives which incorporate one of the following elements: iron, magnesium, sulfur, phosphorous, copper, zinc or selenium.

Iron derivatives, for example, such as the various types of cytochrome enzymes, are an absolute necessity for the evolution of sentient life-forms, up to and including the primates, the pinnacle of all life-forms.

Other elements are also incorporated as needed as the evolutionary pathways advance.

Then, some of these sentient compounds are assembled into rudimentary capsules, the precursors of numerous, different types of cell structures, the organelles: mitochondria, Golgi body, centriole, flagella, cilia, vacuole, plastid, the nucleus, etc.

These capsules effectuate a continuous series of mergers or inclusions; some of which are simply the ingestion of one type of organelle by another type. This eventually leads to the development of ever larger and more complex assemblages of organelles; in which each of the now many types of organelles present express a unique function necessary for an entire group to be able to organize as a unitary body, the cell.

Further changes in internal and external conditions of the cell will cause additional inclusions and differentiations until a definitive cellular line is produced (algae, bacteria, etc.).

Interspersed throughout the cell are a network of interlinked sentinel-type structures, each composed of bonds which pick-up and then convey changes in gravity (B11), magnetism (B10) and electric fluxes (B9).

The resulting conveyed impulses are received by another type of bond structure (B8); which, in turn, then produces its own continuous signals to chemical systems in selective organelles, causing these to produce endless repetitions of the same reaction within each.

The numerous B8 bond-regulated sites in the various organelles in a cell are then coordinated in their actions by neural-type structures, which deploy B7 type bonds, in order to produce a perfectly functioning and long-lived cell.

These B7 bond structures will then, as a secondary function, coordinate the assemblage of cells into specialized multi-cellular systems; which eventually develop into the heart, lungs, skin, skeleton, etc., of a multicellular organism. This is a parallel of the way the B7 bonds had previously coordinated the assembly of many types of diverse organelles into a single, unified cell.

The further evolvement of the multicellular organisms follows the general pathway: fish, amphibians, reptiles, birds and mammals; with the mammalian-primate eventually reaching the highest rung on the evolutionary ladder of complexity and sophistication.

At this juncture in the epilogue, the [V_E]G and [P_E]G gravitons must be reintroduced. These sole deputies of the Singularity in the existent universe have created from each of their unique substances (G4) their equally unique-substance progeny, the V and the P photons, respectively.

Operating as G10 dimensional units, the [V_E]G gravitons, by their propeller-like mixing action, in concert with the [P_E]G gravitons' anvil-type surface, transform the V and the P photons present in the M1 dimension of the magnetic phase into $(VV)^{(+)1/6}$ [(+) monopoles] and $(PP)^{(-)1/6}$ [(-) monopoles)], respectively (M4).

Then, the two types of gravitons further transform into G11 dimensional units; and again, using similar but modified actions as were used in their G10 mode, cause the conversion of the respective monopoles into polymers: positron polymers in the case of the (+) monopoles, and electron polymers with respect to (-) monopoles (M7). These polymers will then eventually depolymerize into respective positrons and electrons as temperatures decline on the periphery of the M7, as well as the E1, E4 and E7 dimensions.

Although the G11 graviton actions were last cited herein in respect to photonic reactions in the E1, M7 and B5 dimensions, suffice it to say that all of the gravitons have continued to be involved in every aspect of the evolvement of all photonic entities in all successive dimensions thereafter, up to and including E11, M11, N11 and B11 dimensions.

As a direct result, each $[V_E]$G and $[P_E]$G graviton has explored and experienced the novelty of their own substance, as well as the other type of graviton's totally foreign substance, as it shepherded the myriad number of V and P photon combinations through forty-two dimensions over billions of years.

In doing this, the gravitons have also acquired an intimate knowledge and understanding of each of the fifty-five distinctly different dimensions of the existent universe.

However, there was one major aspect of the existent universe which had eluded the gravitons, causing them to still not know themselves completely. This last frontier involved acquiring an integrated state of consciousness of being conscious of the unity or oneness of all fifty-five dimensions and the entities, forces, fields and bonds composing them.

All that each graviton had so far acquired was fifty-five separate and individual snapshots.

The one completely integrated picture and absolute sense of a grand unified fifty-five dimensional existent universe and all of their components, a direct parallel to the unity of the Singularity and the existent universe, was still to be sought out and determined.

In order to accomplish this end, myriad lines of multicellular organisms were evolved and explored, finally culminating in the selection of a photonic entity in a branch of the hominid-primate family, the homo sapiens, as the best vehicle for this undertaking.

During the final evolutionary steps in the evolvement of the hominid body, a structure facilitating the connection between it and the graviton was carefully developed. Once operative, it permitted the graviton to completely control every aspect of the hominid-vehicle.

The actual embedding of the graviton, as well as the finalization of the operating connections between it and the homo sapiens body, takes place exclusively during the final labor or C-section process.

After the expulsion from the womb, the act of taking the first breath by the graviton-homo sapiens entity irreversibly seals their union and initiates their life-cycle. This union can only be terminated by the death process; at the end of which the graviton once more becomes a free unitary body, while the homo sapiens-vehicle is returned to its elements and compounds via interment, etc.

However, prior to the first breath, the graviton may sever its connections with the homo sapiens-foetus, thereby terminating the embedding process, as well as the union between them. In this case, the homo sapiens-foetus is stillborn and the graviton is again free to enter into another union with another homo sapiens-foetus.

The rule in all of the embedding processes entered into is: the $[V_E]$G gravitons seek out and effectuate a union with a male foetus, while the $[P_E]$G gravitons seek out a union with a female foetus.

This embedding and union process between a graviton and a suitable homo sapiens-vehicle will be repeated a myriad number of times by each graviton; each union ended by the death process.

This will continue until each graviton has experienced the ultimate union in which it acquires a consciousness of being conscious of the entire existent universe as a grand unified entity.

Then, no further unions will be required of that graviton.

Eventually, each and every graviton will experience this ultimate union, thus ending the need for any further union.

Consciousness of Being Conscious

The graviton and the homo sapiens-vehicle are each separately and differently sentient, or conscious. Their respective substance and structure are distinctly different, due to the entire photonic substance and structure of the homo sapiens-vehicle having been generated from the graviton's entirely unique substance.

Thus, the spectrum of sentient signals and responses to those signals are vastly different in the graviton guest and homo sapiens-vehicle host, due primarily to the graviton being a G1-11 dimensional entity exclusively, while the homo sapiens-host is absent access to these eleven dimensions and the bond B1-B2; as well as being restricted to its own forty-two dimensions.

However, together they each supplement each other, creating a unique partnership able to explore and evaluate the sentient or conscious nature of the fifty-five dimensional, existent universe.

A secondary factor here which widens the consciousness gulf between guest and its host even further is that the graviton's substance and structure is immortal, while the term of the homo sapiens-vehicle is in the tens of years. The former's consciousness is in terms of ages of development; while the latter can only seek out the short-term, as it is a limited-term creature. The graviton will change photonic bodies an untold number of times in the sweep of the Singularity-to-Singularity epoch.

Since the graviton, during its initial period as a guest within the homo sapiens-host, depends on its host almost exclusively for the pick-up and conveying of sentient signals from the hosts internal and external environment, it will almost always tend to believe that the guest and host are one and the same. The realization that they are two distinctly different and independent entities may escape nearly all of the gravitons during the entire term within the confines of their host; except for those gravitons which have already experienced a great number of previous graviton and homo sapiens-vehicle unions.

These latter types of gravitons, having fully become conscious of their own separate and unique consciousness, as well as the independent source of the consciousness of its homo sapiens-host, are still able to operate in a seamless union with the host in order to fully complete their life-cycle.

These fully-realized gravitons are then ready to repeat the events and processes of the G7 through G1 dimensions, as well as the partition event, but in reverse order; thus once more entering into a union with the oneness and nothingness of the Singularity.

Conclusion

The partition event, initiated within the Singularity, created the existent universe.

One day it will be performed again, but this time in reverse; once again only the Singularity is.

We shall, therefore, summarize all that has been previously presented as follows:

One

there is;

One Law

there is;

One Knowledge

there is;

many paths

from

and

to

One

there is;

INDEX

O

P

Q

R

S

Y

Z